MIX DESIGN METHODS

For Asphalt Concrete and Other Hot-Mix Types

- Cover
- Title Page
- Table of Contents
- CH 5 Marshall Method of Mix Design (55-78)

Asphalt Institute
Manual Series No. 2 (MS-2)
Sixth Edition

ASPHALT INSTITUTE

EXECUTIVE OFFICES AND RESEARCH CENTER
Research Park Drive • P.O. Box 14052
Lexington, KY 40512-4052 USA
Telephone 606-288-4960 • FAX 606-288-4999

MEMBERS OF THE ASPHALT INSTITUTE
(As of September 1997)

The Asphalt Institute is an international, nonprofit association sponsored by members of the petroleum asphalt industry to serve both users and producers of asphalt materials through programs of engineering service, research and education. Membership is available to refiners of asphalt from crude petroleum; to processors manufacturing finished paving asphalts and/or non-paving asphalts but not starting from crude petroleum; and to companies working specifically with asphalt related raw material or asphalt additives.

AMOCO OIL COMPANY, Oak Brook, Illinois
†ANDRIE INC., Muskegon, Michigan
ASHLAND PETROLEUM CO., Ashland, Kentucky
ASHWARREN INTERNATIONAL INC.,
  Vancouver, British Columbia, Canada
ASPHALT MATERIALS, INC., Indianapolis, Indiana
†BASF CORPORATION, Charlotte, North Carolina
BITUMAR INC., Montreal, Quebec, Canada
BITUMINOUS PRODUCTS CO., Maumee, Ohio
†BOCHARD COASTWISE MANAGEMENT CORP.,
  Hicksville, New York
CALTEX PETROLEUM CORPORATION, Dallas, Texas
CANADIAN ASPHALT INDUSTRIES INC.,
  Markham, Ontario, Canada
†CHEMEX, INC., Bakersfield, California
CITGO ASPHALT REFINING COMPANY,
  Blue Bell, Pennsylvania
†COASTAL TOWING, INC., Houston, Texas
COMPANIA ESPANOLA DE PETROLEOS, S.A.,
  Madrid, Spain
CONOCO INC., Houston, Texas
CONSOLIDATED OIL & TRANSPORTATION
  COMPANY, INC., Englewood, Colorado
ECOPETROL - ICP, Santander, Colombia
EGYPTIAN GENERAL PETROLEUM CORP., THE,
  Cairo, Egypt
†ENICHEM ELASTOMERS AMERICAS, INC., Houston,
  Texas
EXXON COMPANY, U.S.A., Houston, Texas
FINA OIL AND CHEMICAL COMPANY, Dallas, Texas
FRONTIER TERMINAL & TRADING COMPANY, Tulsa,
  Oklahoma
GOLDEN BEAR OIL SPECIALTIES, Los Angeles,
  California
†HEATEC, INC., Chattanooga, Tennessee
†HOLLYWOOD MARINE, INC., Houston, Texas
HUNT REFINING CO., Tuscaloosa, Alabama
IMPERIAL OIL, Toronto, Ontario, Canada
JEBRO, INC, Sioux City, Iowa
KOCH MATERIALS COMPANY, Wichita, Kansas
KOCH MATERIALS LTD., Toronto, Ontario, Canada
  A Subsidiary of Koch Oil Company Ltd.
MARATHON OIL COMPANY, Findlay, Ohio
MATHY CONSTRUCTION COMPANY, Onalaska,
  Wisconsin
MCASPHALT INDUSTRIES LTD.,
  Scarborough, Ontario, Canada

MIDDLEPORT TERMINAL INC., Gallipolis, Ohio
MOBIL OIL CORPORATION, Fairfax, Virginia
†MORANIA OIL TANKER CORP., Stamford,
  Connecticut
MURPHY OIL USA, INC., Superior, Wisconson
NAVAJO REFINING COMPANY, Artesia, New Mexico
NESTE TRIFINERY PETROLEUM SERVICES,
  Houston, Texas
NYNÄS PETROLEUM, AB, Johanneshov, Sweden
PARAMOUNT PETROLEUM CORPORATION,
  Paramount, California
PETRO-CANADA INC., North York, Ontario,
  Canada
PETROLEO BRASILEIRO S.A. - PETROBRAS,
  Rio de Janeiro, Brazil
REFINADORA COSTARRICENSE DE PETROLEO
  (RECOPE), San Jose, Costa Rica
REFINERIA DE PETROLEO CONCON S.A., Concon,
  Chile
REFINERIA SAN LORENZO, S.A., San Lorenzo,
  Argentina
REPSOL PRODUCTOS ASFALTICOS, S.A., Madrid,
  Spain
†SAFETY-KLEEN, OIL RECOVERY DIVISION,
  Breslau, Ontario, Canada
SAN JOAQUIN REFINING CO., INC., Bakersfield,
  California
SARGEANT MARINE, INC., Coral Gables, Florida
SENECA PETROLEUM CO., INC., Crestwood, Illinois
SHELL CANADA LIMITED, Toronto, Ontario, Canada
SHELL INTERNATIONAL PETROLEUM COMPANY
  LIMITED, London, England
SHELL OIL PRODUCTS COMPANY, Houston, Texas
SOUTHLAND OIL COMPANY, Jackson, Mississippi
T-M OIL COMPANY, INC., Belleville, Michigan
TIPCO ASPHALT PUBLIC COMPANY LIMITED,
  Bangkok, Thailand
TRUMBULL PRODUCTS, Toledo, Ohio
  Division of Owens Corning
†ULTRAPAVE DIVISION, TEXTILE RUBBER &
  CHEMICAL CO., Dalton, Georgia
UNITED REFINING CO., Warren, Pennsylvania
YPF S.A., Buenos Aires, Argentina
YUKONG LIMITED, Ulsan, Korea

† Affiliate Member

Foreword

This manual is a practical guide to asphalt mix design for engineers and others concerned with the technicalities of constructing all types of pavement with hot mix asphalt. It also serves as an excellent textbook for students being initially exposed to asphalt mix design.

This sixth edition of the *Mix Design Manual* includes these revisions from previous editions: the addition of a Voids Filled with Asphalt (VFA) criterion to Marshall mix design; the recommendation to initially select asphalt content at four percent air voids; the redefinition of the nominal maximum aggregate size; discussions to assist the designer in making minor changes in the designed mix; mix design using reclaimed asphalt pavement (RAP); and procedures used in field verification of asphalt mixtures.

At the time this sixth edition of the Mix Design Manual was released, the asphalt mix design procedures being developed in the Strategic Highway Research Program (SHRP) had not been finalized. Therefore, the SHRP mix design procedures have not been included here. Please contact the Asphalt Institute for the most recent information concerning SHRP mix design.

Asphalt Institute
Lexington, Kentucky

ASPHALT INSTITUTE
EXECUTIVE OFFICES AND RESEARCH CENTER
Research Park Drive • P.O. Box 14052
Lexington, KY 40512-4052 USA
Telephone 606-288-4960 • FAX No. 606-288-4999

ASPHALT INSTITUTE ENGINEERING OFFICES
(As of March 1995)

EASTERN REGION

*NORTH LITTLE ROCK, AR 72190-4007—P.O. Box 4007 (5318 John F. Kennedy
Blvd.), (501) 758-0484, Arkansas, Kansas, Missouri, Eastern Oklahoma

RALEIGH, NC 27605—2016 Cameron Street, Suite 208, (919) 828-5998,
North Carolina, South Carolina, Virginia

BETHESDA, MD 20814—6917 Arlington Road, Suite 210, (301) 656-5824
District of Columbia

COLUMBUS, OH 43232—2238 South Hamilton Road, Suite 103, (614) 759-1400,
Ohio, Kentucky, West Virginia

DILLSBURG, PA 17019—P.O. Box 337, (2-4 E. Harrisburg Street), (717) 432-5965
Pennsylvania, Delaware, Maryland

METHUEN, MA 01844—248 Pleasant Street, Room 204, (508) 681-0455,
Connecticut, Maine, Massachusetts, New Hampshire, Rhode Island, Vermont

PANAMA CITY, FL 32405—2639-B Lisenby Avenue, (904) 763-3363,
Florida, Georgia, Alabama

TRENTON, NJ 08611—527 Chestnut Avenue, (609) 393-1466, New York, New
Jersey

SAN ANTONIO, TX 78233—10635 IH-35N, Suite 101, (210) 590-9644,
Texas (except Southwest Texas), Western Oklahoma

CHICAGO, IL 60643—P.O. Box 439081, (708) 388-2001,
Illinois, Wisconsin, Iowa

INDIANAPOLIS, IN 46268—P.O. Box 68463 (5455 West 86th Street, Suite 113),
(317) 872-1412, Indiana, Michigan

JACKSON, MS 39206—112 Office Park Plaza, Suite 13, (601) 981-3417,
Louisiana, Mississippi, Tennessee

WESTERN REGION

*AUBURN, CA 95603—164 Maple Street, No. 3A, (916) 885-2625,
Northern California, Northern Nevada, Hawaii

LITTLETON, CO 80122—7000 South Broadway, Suite 2B, (303) 798-2972,
Colorado, Montana, Wyoming

OLYMPIA, WA 98502—2626 12th Court, S.W., (360) 786-5119,
Oregon, Idaho, Washington, Alaska

TEMPE, AZ 85282—201 East Southern Avenue, Suite 118, (602) 829-0448,
Arizona, New Mexico, Southwest Texas, Utah

WEST LAKE VILLAGE, CA 91362—3609 Thousand Oaks Boulevard, Suite 216,
(805) 373-5130, Southern California, Southern Nevada

ST. CLOUD, MN 56302-0941—P.O. Box 941 (921 First St. North) (612) 654-0744,
Minnesota, North Dakota, South Dakota, Nebraska

*Regional Office

CONTENTS

ILLUSTRATIONS

TABLES

Chapter 1

Introduction

1.01 PURPOSE AND SCOPE — The objective of hot mix asphalt (HMA) mix design is to determine the combination of asphalt cement and aggregate that will give long lasting performance as part of the pavement structure. Mix design involves laboratory procedures developed to establish the necessary proportion of materials for use in the HMA. These procedures include determining an appropriate blend of aggregate sources to produce a proper gradation of mineral aggregate, and selecting the type and amount of asphalt cement to be used as the binder for that gradation. Well-designed asphalt mixtures can be expected to serve successfully for many years under a variety of loading and environmental conditions.

The mix design of hot mix asphalt is just the starting point to assure that an asphalt concrete pavement layer will perform as required. Together with proper construction practice, mix design is an important step in achieving well-performing asphalt pavements. In many cases, the cause of poorly-performing pavements has been attributed to poor or inappropriate mix design or to the production of a mixture different from what was designed in the laboratory. Correct mix design involves adhering to an established set of laboratory techniques and design criteria. These techniques and criteria serve as the design philosophy of the governing agency. They are based on scientific research as well as many years of experience in observing the performance of asphalt concrete pavements. It is critical that these laboratory methods be followed exactly as written.

Successful mix design requires understanding the basic theory behind the steps and following the intent of the written instructions. It also includes having the proper training in laboratory techniques and effectively interpreting the results of laboratory tests. This manual was prepared with these goals in mind. It contains the latest information for the design of hot-mix asphalt paving mixtures to meet the demands of modern traffic conditions and to ensure optimal performance of asphalt concrete pavements.

Chapter 2 of this manual relates the application of mix design and testing to general practice. Testing references and detailed procedures are outlined for the routine analysis of materials and paving mixtures. A number of aggregate gradation computations with typical examples of routine calculations related to mix design are included in Chapter 3. Chapter 4 describes the asphalt mixture properties important to the long term performance of asphalt pavements.

The principal features of this manual are the detailed presentations for two methods of asphalt paving mix design (Marshall Method in Chapter 5 and Hveem Method in Chapter 6). The test procedures for each mix design method are described, along with the corresponding guidelines and procedures for selecting the design asphalt content. Many of these calculations and guidelines are included in the Asphalt Institute

Computer-Assisted Asphalt Mix Analysis (CAMA) computer program. The Appendix presents the addition of reclaimed asphalt pavement (RAP) into Marshall and Hveem mix design.

Each mix design method and the corresponding test criteria are presented without any specification requirements for materials and construction. The compaction method and the level of compaction energy approximate the degree of compaction that will exist in the pavement after several years of traffic. The design asphalt content is chosen to provide for all of the mix components (asphalt, aggregate, and air) to be in correct proportion at this point in time.

The Marshall and Hveem methods of mix design are both widely used for the design of hot mix asphalt. The selection and use of either of these mix design methods is principally a matter of engineering preference, since each method has certain unique features and apparent advantages. Both methods are currently being used with satisfactory results when all of the principles of proper mix analysis are observed.

The durability of aggregates and asphalt-aggregate compatibility can be a major concern in some cases. Additional material testing topics are covered in Chapter 7.

As stated earlier, laboratory mix design is just the starting point of the process. To ensure that the mix being placed in the pavement is the same as the mix designed and evaluated in the lab, field verification and quality control are essential. Chapter 8 describes the various facets of quality management systems for asphalt mixes.

1.02 HOT MIX DEFINED — Hot mix asphalt paving materials consist of a combination of aggregates that are uniformly mixed and coated with asphalt cement. To dry the aggregates and obtain sufficient fluidity of the asphalt cement for proper mixing and workability, both must be heated prior to mixing—giving origin to the term "hot-mix."

The aggregates and asphalt are combined in an asphalt mixing facility, continuously or in batch-mode. These two main components are heated to proper temperature, proportioned, and mixed to produce the desired paving material. After the plant mixing is complete, the hot-mix is transported to the paving site and spread with a paving machine in a partially-compacted layer to a uniform, smooth surface. While the paving mixture is still hot, it is further compacted by heavy self-propelled rollers to produce a smooth, well-consolidated course of asphalt concrete.

1.03 CLASSIFICATION OF HOT MIX ASPHALT PAVING — Asphalt paving mixes may be designed and produced from a wide range of aggregate blends, each suited to specific uses. The aggregate composition typically varies in size from coarse to fine particles. Many different compositions are specified throughout the world — the mixes designated in any given locality generally are those that have proven adequate through long-term usage and, in most cases, these gradings should be used.

For a general classification of mix compositions, the Asphalt Institute recommends consideration of mix designations and nominal maximum size of aggregate: 37.5 mm (1-1/2 in.), 25.0 mm (1 in.), 19.0 mm (3/4 in.), 12.5 mm (1/2 in.), 9.5 mm (3/8 in.), 4.75 mm (No. 4), and 1.18 mm (No. 16), as specified in the American Society for Testing and Materials (ASTM) Standard Specification D 3515 for *Hot-Mixed, Hot-Laid*

Bituminous Paving Mixtures. The grading ranges and asphalt content limits of these uniformly-graded dense mixes generally agree with overall practice but may vary from the practice of a particular local area. Further discussion of asphalt mixture gradations is presented in Article 2.03.

Depending on the specific purpose of the mix, other non-uniform gradings have been used with great success, such as gap-graded and open-graded aggregate compositions. The design philosophy and construction procedures of these mixes are different because of the additional void space incorporated between the larger particles. The design procedures in this manual should not be used for gap-graded or open-graded asphalt mixtures.

Chapter 2

Mix Design Practice

2.01 GENERAL — Asphalt paving mix design <u>demands attention</u> to the details outlined in standard test procedures. Primarily, this means following specific, written instructions. But it also means having proper training in laboratory technique and the relation of mix design testing to pavement field specification requirements.

While mix design often is treated as an isolated subject, it cannot be separated from the other related items of the material specifications. It is the purpose of this chapter, therefore, to cite the general objectives of mix design and present a guide for applying the mix design principles to asphalt paving construction specifications.

2.02 OBJECTIVES OF ASPHALT PAVING MIX DESIGN — The design of asphalt paving mixes, as with other engineering materials designs, is largely a matter of selecting and proportioning materials to obtain the desired properties in the finished construction product. *The overall objective for the design of asphalt paving mixes is* to determine (within the limits of the project specifications) a *cost-effective blend and gradation of aggregates and asphalt* that yields a mix having:

(1) *Sufficient asphalt* to ensure a durable pavement.
(2) *Sufficient mix stability* to satisfy the demands of traffic without distortion or displacement.
(3) *Sufficient voids* in the total compacted mix to allow for a slight amount of additional compaction under traffic loading and a slight amount of asphalt expansion due to temperature increases without flushing, bleeding, and loss of stability.
(4) *A maximum void content* to limit the permeability of harmful air and moisture into the mix.
(5) *Sufficient workability* to permit efficient placement of the mix without segregation and without sacrificing stability and performance.
(6) For surface mixes, *proper aggregate texture* and *hardness* to provide sufficient skid resistance in unfavorable weather conditions.

The final goal of mix design is to select a unique design asphalt content that will achieve a balance among all of the desired properties. Ultimate pavement performance is related to durability, impermeability, strength, stability, stiffness, flexibility, fatigue resistance, and workability. Within this context, there is no single asphalt content that will maximize all of these properties. Instead, an asphalt content is selected on the basis of optimizing the properties necessary for the specific conditions.

Since the fundamental performance properties are not directly measured in a normal mix design, asphalt content is selected on the basis of a measured parameter that best reflects all of these desires. Considerable research has determined that air void content is this parameter. An acceptable air voids range of three to five percent is most often used. Within this range, four percent air voids is often considered the best initial estimate for a design that balances the desired performance properties. Slight refinements are then considered in the analysis of the mix testing results.

Mix Type Selection

2.03 GENERAL — Dense-graded HMA mixtures are generally divided into three major categories dependent upon their specific use: surface mixtures, binder or intermediate mixtures, and base mixtures. HMA mixtures are typically designed with layer thickness and availability of aggregates in mind. The maximum size aggregate is generally largest in the base, smaller in the binder or intermediate course, and finest in the surface course; however, this practice is not universal. Nevertheless, any properly designed HMA mix can generally serve at any level in the pavement. Surface course mixtures may become "binder" mixes if subsequently overlaid, so strength requirements should not be compromised regardless of the location of the mix within the pavement.

Generally, there is no single, uniform standard set of HMA classifications used by the various public agencies. There are similarities with respect to mixture types, but the geographic availability of materials and different climatic design requirements have led to various identifications. Each agency usually has its own designation for identifying various mixture types. While most HMA mixtures have a typical design use, these mixes offer a wide range of performance characteristics and there is substantial overlap of mixture application.

This article describes the various types of HMA mixtures and typical applications. One national standard that identifies HMA according to maximum aggregate size and gradation is ASTM D 3515, *Standard Specifications for Hot-Mixed, Hot-Laid Bituminous Paving Mixtures*. The aggregate gradations given in the various figures have been taken from this specification. Table 2.1 presents the dense-graded mixture gradations from ASTM D3515. HMA mix types can generally be narrowed down to discussions of the mixture gradation (dense-graded or open-graded) and the maximum aggregate size (sand-asphalt up to "large-stone" mixes).

Depending on the gradation, pavement layers are confined to practical minimum and maximum lift thicknesses. The minimum thickness for a surface mix usually varies from 2 to 3 times the maximum aggregate size; however, the actual minimum thickness of any course is that which can be demonstrated to be laid in a single lift and compacted to the required uniform density and smoothness. The maximum lift thickness is usually governed by the ability of the rollers to achieve the specified compaction for that layer.

Regardless of the mixture classification, the same degree of design, production, and construction control procedures should be used to ensure proper performance of the pavement.

Surface Course Mixtures

Surface course mixes must be designed to have sufficient stability and durability to both carry the anticipated traffic loads and to withstand the detrimental effects of air, water, and temperature changes. In general, surface mixtures have a greater asphalt content than binder or base mixtures due to the higher VMA requirements of smaller maximum aggregate size mixtures. Maximum aggregate sizes for surface mixes vary from 9.5 to 19 mm (3/8 to 3/4 in.). The choice of maximum aggregate size is often

Table 2.1 – Composition of asphalt paving mixtures (ASTM D3515)

Dense Mixtures

Mix Designation and Nominal Maximum Size of Aggregate

Grading of Total Aggregate (Coarse Plus Fine, Plus Filler if Required)
Amounts Finer Than Each Laboratory Sieve (Square Opening), Weight %

Sieve Size	2 in. (50 mm)	1 1/2 in. (37.5 mm)	1 in. (25.0 mm)	3/4 in. (19.0 mm)	1/2 in. (12.5 mm)	3/8 in. (9.5 mm)	No. 4 (4.75 mm) (Sand Asphalt)	No. 8 (2.36 mm)	No. 16 (1.18 mm) (Sheet Asphalt)
2 1/2 in. (63-mm)	100								
2 in. (50-mm)	90 to 100	100							
1 1/2 in. (37.5-mm)	60 to 80	90 to 100	100						
1 in. (25.0-mm)			90 to 100	100					
3/4 in. (19.0-mm)	35 to 65	56 to 80	56 to 80	90 to 100	100				
1/2 in. (12.5-mm)					90 to 100	100			
3/8 in. (9.5-mm)				56 to 80		90 to 100	100		
No. 4 (4.75-mm)[A]	17 to 47	23 to 53	29 to 59	35 to 65	44 to 74	55 to 85	80 to 100		100
No. 8 (2.36-mm)[A]	10 to 36	15 to 41	19 to 45	23 to 49	28 to 58	32 to 67	65 to 100		95 to 100
No. 16 (1.18-mm)							40 to 80		85 to 100
No. 30 (600-μm)	3 to 15	4 to 16	5 to 17	5 to 19	5 to 21	7 to 23	25 to 65		70 to 95
No. 50 (300-μm)							7 to 40		45 to 75
No. 100 (150-μm)[B]							3 to 20		20 to 40
No. 200 (75-μm)[B]	0 to 5	0 to 6	1 to 7	2 to 8	2 to 10	2 to 10	2 to 10		9 to 20

Bitumen, Weight % of Total Mixture[C]

Sieve Size	2 in. (50 mm)	1 1/2 in. (37.5 mm)	1 in. (25.0 mm)	3/4 in. (19.0 mm)	1/2 in. (12.5 mm)	3/8 in. (9.5 mm)	No. 4 (4.75 mm)	No. 8 (2.36 mm)	No. 16 (1.18 mm)
	2 to 7	3 to 8	3 to 9	4 to 10	4 to 11	5 to 12	6 to 12	7 to 12	8 to 12

Suggested Coarse Aggregate Sizes

Sieve Size	2 in. (50 mm)	1 1/2 in. (37.5 mm)	1 in. (25.0 mm)	3/4 in. (19.0 mm)	1/2 in. (12.5 mm)	3/8 in. (9.5 mm)	No. 4 (4.75 mm)	No. 8 (2.36 mm)	No. 16 (1.18 mm)
	3 and 57	4 and 67 or 4 and 68	5 and 7 or 57	67 or 68 or 6 and 8	7 or 78	8			

A In considering the total grading characteristics of a bituminous paving mixture, the amount passing the No. 8 (2.36-mm) sieve is a significant and convenient field control point between the fine and coarse aggregate. Gradings approaching the maximum amount permitted to pass the No. 8 sieve will result in pavement surfaces having comparatively fine texture, while coarse gradings approaching the minimum amount passing the No. 8 sieve will result in surfaces with comparatively coarse texture.

B The material passing the No. 200 (75-μm) sieve may consist of fine particles of the aggregates or mineral filler, or both but shall be free of organic matter and clay particles. The blend of aggregates and filler, when tested in accordance with Test Method D 4318, shall have a plasticity index of not greater than 4, except that this plasticity requirement shall not apply when the filler material is hydrated lime or hydraulic cement.

C The quantity of bitumen is given in terms of weight % of the total mixture. The wide difference in the specific gravity of various aggregates, as well as a considerable difference in absorption, results in a comparatively wide range in the limiting amount of bitumen specified. The amount of bitumen required for a given mixture should be determined by appropriate laboratory testing or on the basis of past experience with similar mixtures, or by a combination of both.

The American Society for Testing and Materials takes no position respecting the validity of any patent rights asserted in connection with any item mentioned in this standard. Users of this standard are expressly advised that determination of the validity of any such patent rights, and the risk of infringement of such rights, are entirely their own responsibility.

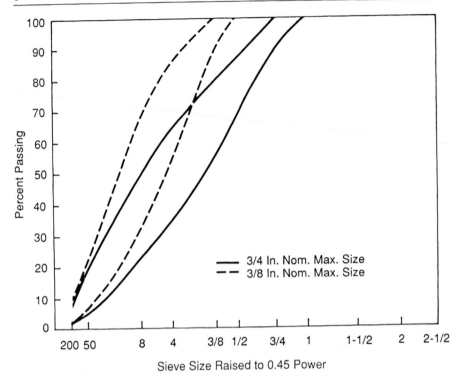

Figure 2.1 – **Typical surface course gradations**

predicated on the desired surface texture, with a smaller maximum size aggregate producing a smoother, tighter surface. Figure 2.1 illustrates typical gradation ranges of 9.5mm (3/8 in.) and 19 mm (3/4 in.) nominal maximum size dense-graded mixtures.

A special type of surface mixture used for reducing hydroplaning and increasing skid resistance is an open-graded friction course (OGFC), also known as a porous friction course (PFC) or popcorn mix. The function of this mixture is to provide a free-draining layer that permits surface water to migrate laterally through the mixture to the edge of the pavement. The open-graded mixture also provides a skid resistant surface as its coarse texture provides excellent friction between the pavement and the tire. OGFCs contain a relatively high asphalt content using a 9.5 to 12 mm (3/8 to 1/2 in.) maximum size aggregate, with few aggregate fines to produce the open-graded mixture. Typically placed in 16 to 19 mm (5/8 to 3/4 in.) thicknesses, the mixes are placed only to facilitate rapid removal of surface water and not as an improvement to structural capacity. Figure 2.2 shows the gradation range of a typical 9.5 mm (3/8 in.) open-graded friction course.

Binder Course Mixtures

Binder mixes are often used as an intermediate layer between the surface mixture and the underlying asphalt or granular base. Binder mixes typically have a larger maximum size aggregate of 19 to 38 mm (3/4 to 1.5 in.), with a corresponding lower

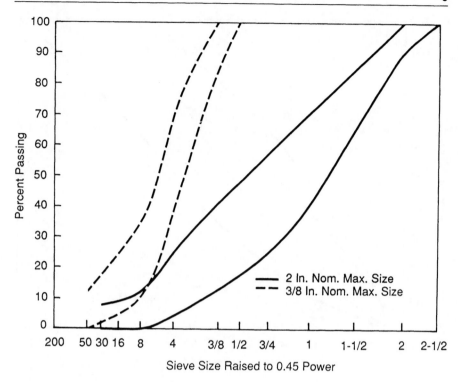

Figure 2.2 – **Typical open-graded mixture gradations**

asphalt content. Binder and base mixes are often used interchangeably in pavement design and construction. Where heavy wheel loads are involved, a typical binder mix for highway construction can be used as a surface mix if a coarser surface texture will not be a concern. This approach has often been used in port facilities using heavy cargo handling vehicles; in logging yards that use large log-handling vehicles; and for truck unloading and industrial areas with high percentages of heavy trucks. Larger aggregate mixes (with less asphalt and sand contents) are often more resistant to the scuffing action of tight radius power steering turns. Figure 2.3 shows the gradation range of a 25 mm (1-in.) nominal maximum size dense-graded mix.

Base Course Mixtures

Hot mix asphalt base mixes can be placed directly on the compacted subgrade or over a granular base. HMA base mixes are characterized by larger aggregate sizes that range up to 75 mm (3 in.). The relative asphalt content will be lower due to the larger maximum aggregate size, which is appropriate because this mixture is not exposed to climatic factors. Maximum aggregate sizes for base mixtures are often established by the locally available material. Figure 2.3 illustrates the gradation range of a 50 mm (2-in.) nominal maximum size dense-graded HMA.

Base mixes can also be designed with an open gradation to facilitate drainage of water that may eventually enter the pavement structure. A similar type open-graded

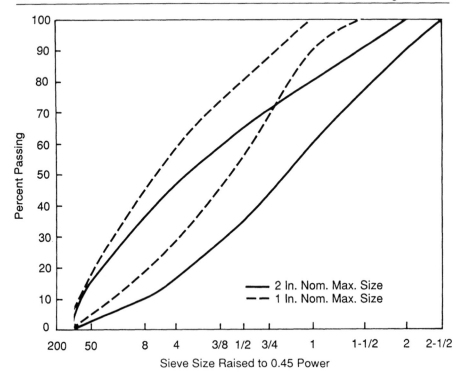

Figure 2.3 – **Typical binder and base course gradations**

mix is used as a crack-relief layer in pavement rehabilitation. Open-graded base mixtures are designed to provide an interconnecting void structure, using 100 percent crushed materials with maximum aggregate sizes of about 38 to 76 mm (1.5 to 3 in.). Positive, free drainage must be incorporated in the overall pavement design with these layers. Figure 2.2 shows the gradation range of a typical 50 mm (2-in.) open-graded mixture.

Sand-Asphalt Mixes
An appropriately graded manufactured sand or natural sand or a combination of both can be used effectively as either a base or surface mixture. The primary difference between a base and surface sand-asphalt mix would be in the amount of asphalt cement and minus 75 μ m (No. 200) material that may be specified. Also known in some areas as a plant-mix seal or as sheet asphalt, sand-asphalt mixes do not normally have the high stability associated with larger-sized aggregate mixtures. These types of mixtures are not recommended in heavy traffic-load areas.

Sand-asphalt mixes produce the tightest surface texture of any HMA and with proper selection of aggregate type (hardness and shape) can also produce a highly skid resistant mixture. An additional advantage of sand mixes is that they can be placed in thicknesses as thin as 15 mm (0.60 in.). For this reason sand mixes can be used as a thin leveling course prior to an HMA overlay. A sand asphalt mix can be made into a sheet asphalt by the addition of relatively large amounts of mineral filler and asphalt cement.

2.04 DESIGN METHOD AND REQUIREMENTS — Ultimately, one should recognize that the mix design method and design requirements form an *essential* part of the construction specifications for asphalt paving projects. The construction agency or authority responsible for of the paving construction usually establishes the mix design method and design requirements. Once these items are established, it then becomes the duty and responsibility of the engineer to do the mix design within the framework of all the specification requirements.

The Marshall and Hveem methods of mix design presented in this manual have been widely used with satisfactory results. For each of these methods, criteria have been empirically developed by correlating the results of laboratory tests on the compacted paving mixes with the performance of the paving mixes under service conditions. In each instance, however, the correlation was made within certain limits; these limits are clearly listed for each method.

IMPORTANT: For the above reasons, the design criteria for each mix design method are applicable only to the prescribed test procedure within the limits of the original correlation. Hasty or haphazard modification of these design methods, test procedures or design criteria is never justified. In those cases where it can be clearly shown that a modification or extension of the design method is needed, all proposed changes should be fully supported with additional correlation data covering the new limits or conditions of design.

All mix design procedures involve preparing a set of trial mixture specimens using materials proposed for use on the project. An examination of the standard procedures will indicate that there are three key components of mix design:

- laboratory compaction of trial mix specimens,
- stability (or strength) and volumetric testing, and
- analysis of results.

An additional step that is becoming more common is the evaluation of moisture susceptibility or the compatibility of the aggregate and the asphalt cement.

The purpose of laboratory compaction is not to produce conveniently-sized trial mixture specimens. The compaction technique is intended to simulate the in-place density of HMA after it has endured several years of traffic. Numerous studies have been done to compare the measured properties of cored specimens to laboratory-compacted specimens of actual plant-mixed materials. Research has failed to establish one compaction method which consistently produces the closest simulation to the field for <u>all</u> of the measured properties. Four compaction methods are currently in use:

- impact compaction, used in the Marshall mix design method
- kneading compaction, used in the Hveem mix design method
- several forms of gyratory compaction
- compaction using vibratory impact hammers

Various agencies have investigated the use of gyratory compaction to replace the currently accepted compaction procedures. The impact and kneading compaction procedures used in the Marshall and Hveem mix design methods, respectively, should not be altered unless extensive studies have been performed to validate the alternate compaction procedure.

After laboratory compaction, the specimens are tested for strength and density or void properties. Normally, five trial mixtures are compacted. The aggregate gradation remains the same for each trial and the asphalt content is varied. To ensure statistically significant results, at least three replicate specimens of each trial mixture (i.e. asphalt content) should be prepared.

2.05 EVALUATION AND ADJUSTMENT OF MIX DESIGNS — When developing a specific mix design, it is often necessary to make several trial mixes to find one that meets all of the design criteria. Each trial mix design, therefore, serves as a guide for evaluating and adjusting the trials that follow. For preliminary or exploratory mix designs it is advisable to start with a blended aggregate gradation that approaches the median of the specification limits. Initial trial mixes are used to establish the job-mix formula and verify that an aggregate gradation within the specification limits can be produced by the central mixing facility. This assurance is particularly important when there are no service records available on the prospective aggregate sources.

When scheduling preliminary mix designs, it should be verified that both asphalt and aggregate materials meet the proposed specification requirements. When several possible sources of aggregate are to be considered, it may be necessary to make a number of trial mix designs to determine the most economical combination of aggregates that will fulfill all of the specification requirements. The results of the preliminary mix designs serve as a basis for making a preliminary estimate of costs.

When the initial trial mixes fail to meet the design criteria at any selected asphalt content, it will be necessary to modify or, in some cases, redesign the mix. To correct a deficiency, the easiest way to redesign a mixture is to change the aggregate gradation by adjusting the component percentages. Often this adjustment is enough to bring all properties within compliance. If adjusting the percentages is not adequate to meet all of the design criteria, serious consideration should be given to changing one or more material sources.

For many engineering materials, the strength of the material frequently denotes quality; however, this is not necessarily the case for hot mix asphalt. Extremely high stability often is obtained at the expense of lowered durability, and vice versa. Therefore, in evaluating and adjusting mix designs always keep in mind that the aggregate gradation and asphalt content in the final mix design must strike a favorable balance between the stability and durability requirements for the use intended.

Moreover, the mix must be produced as a practical and economical construction operation.

Grading curves are helpful in making necessary adjustments in mix designs. For example, curves determined from the Fuller equation* represent mix conditions of maximum density and minimum voids in mineral aggregate (VMA). Paving mixtures with such curves may be easily compacted, but tend to pack very tight and have air void contents that are too low. Usually, deviations from these curves will result in higher VMA and lower densities for the same compactive effort. The extent of change in density and VMA depends on the amount of adjustment in fine or coarse aggregate. Figure 2.4 illustrates a series of Fuller maximum density curves plotted on a conventional semi-log grading chart.

Figure 2.5 illustrates maximum density curves determined from the maximum density equation with particle sizes raised to the 0.45 power** and plotted on the Federal Highway Administration grading chart (based on a scale raising sieve openings to the 0.45 power). Many designers find this chart more convenient to use for adjusting aggregate gradings. The curves on this chart, however, need not be determined from the maximum density equation. They may be obtained by drawing a straight line from the origin at the lower left of the chart to the desired maximum particle size*** at the top 100 percent passing line. Gradings that closely approach this straight line usually must be adjusted away from it within acceptable limits to increase the VMA values. This allows enough asphalt to be used to obtain maximum durability without the mixture flushing.

These are general guidelines for adjusting the trial mix, but the suggestions outlined may not necessarily apply in all cases:

(a) *Voids Low, Stability Low*—Voids may be increased in a number of ways. As a general approach to obtaining higher voids in the mineral aggregate (and therefore providing sufficient void space for an adequate amount of asphalt and air voids) the aggregate grading should be adjusted by adding more coarse or more fine aggregate.

If the asphalt content is higher than normal and the excess is not required to replace that absorbed by the aggregate, the asphalt content may be lowered to increase the voids provided adequate VMA is retained. It must be remembered, however, that lowering the asphalt content may decrease the durability of the pavement. Too much reduction in asphalt content may lead to brittleness, accelerated oxidation, and increased permeability. If the above adjustments do not produce a stable mix, the aggregate may have to be changed.

---

*p = $100(d/D)^{0.5}$ in which, p = total percentage passing given sieve
$\qquad\qquad\qquad\qquad\qquad$ d = size of sieve opening
$\qquad\qquad\qquad\qquad\qquad$ D = largest size (sieve opening) in gradation

**P = $100(d/D)^{0.45}$

***For processed aggregate, the maximum particle size in a standard set of sieves listed in the applicable specification is two sizes larger than the first sieve to retain more than 10 percent of the material.

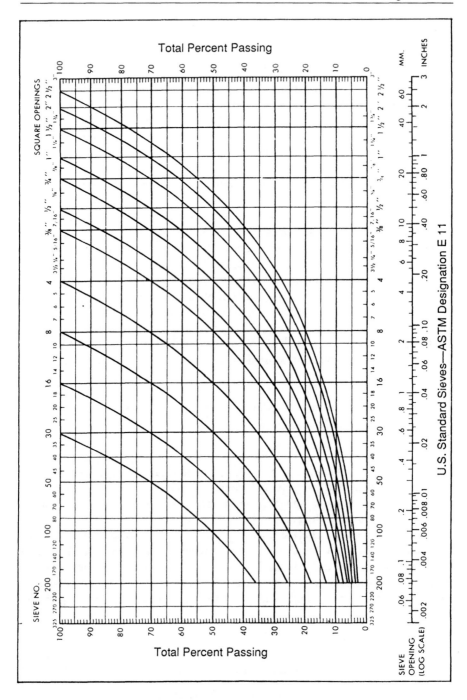

Figure 2.4 – **Fuller maximum density curves on standard semi-log grading chart**

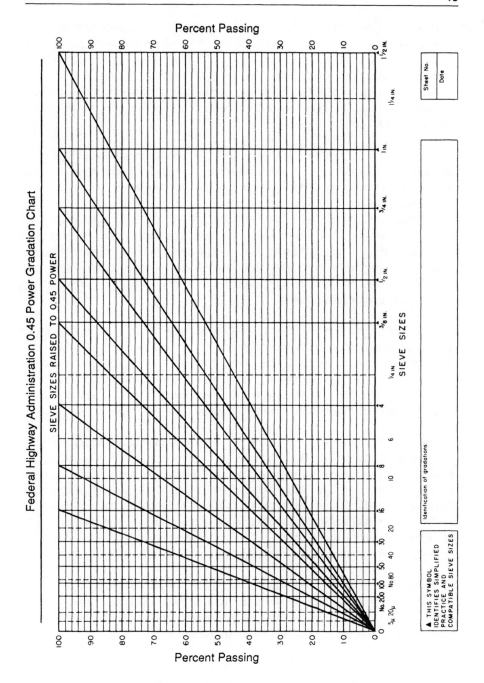

Figure 2.5 – **Maximum density curves on Federal Highway Administration 0.45 power gradation chart**

It usually is possible to improve the stability and increase the aggregate void content of the mix by increasing the amount of crushed materials and/or decreasing the amount of material passing the 75 μ m (No. 200) sieve. With some aggregates, however, the freshly-fractured faces are as smooth as the water-worn faces and an appreciable increase in stability is not possible. This is generally true of quartz or similar rock types. By adding more manufactured sand, the void content can also be improved without sacrificing mix stability.

(b) *Voids Low, Stability Satisfactory*—Low void content may eventually result in instability due to plastic flow or flushing after the pavement has been exposed to traffic for a period of time because of particle reorientation and additional compaction. Insufficient void space may also result because of the amount of asphalt required to obtain high durability in finer mixes, even though stability is initially satisfactory for the specific traffic. Degradation of a poor aggregate during mixture production and/or under the action of traffic may also subsequently lead to instability and flushing if the void content of the mix is not sufficient. For these reasons, mixes low in voids should be adjusted by one of the methods given in (a) above, even though the stability may initially appear satisfactory.

(c) *Voids Satisfactory, Stability Low*—Low stability when voids and aggregate grading are satisfactory may indicate some deficiencies in the aggregate. Consideration should be given to improving the coarse aggregate particle shape by crushing, or increasing the percentage of coarse aggregate in the mixture, or possibly increasing the maximum aggregate size. Aggregate particles with rougher texture and less rounded surfaces will exhibit more stability while maintaining or increasing the void content.

(d) *Voids High, Stability Satisfactory*—High void contents are frequently associated with mixes found to have high permeability. High permeability, by permitting circulation of air and water through the pavement, may lead to premature hardening of the asphalt, ravelling of aggregate, or possibly stripping of the asphalt off the aggregate. Even though stabilities are satisfactory, adjustments should be made to reduce the voids. Small reductions may be accomplished by increasing the mineral dust content of the mix. It may be necessary to select or combine aggregates to a gradation which is closer to the maximum density grading curve.

(e) *Voids High, Stability Low*—Two steps may be necessary when the voids are high and the stability is low. First the voids are adjusted by the methods discussed above. If this adjustment does not also improve the stability, the second step should be a consideration of aggregate quality as discussed in (a) and (b) above.

2.06 QUALITY MANAGEMENT TESTING SYSTEM — Mix design testing for asphalt paving construction is only the starting point of the process of producing a quality pavement. Field verification testing must be continually performed on the field-produced mixture to ensure that the criteria established and used in laboratory design for the particular mixture are being met on the job. Significant equipment and material differences exist between the small scale operation of the laboratory mixing bowl and an asphalt mixing facility. Field verification of hot mix asphalt is necessary to measure what differences exist and to determine what, if any, corrective measures need to be taken. It is important to note that mix design criteria apply equally to both field produced mixtures and laboratory mixed specimens.

Normally, a total quality management system will have four important phases within the overall project: pre-production sampling and testing, initial job-mix formula verification, daily job-mix control testing during production, and in-place acceptance testing. The following outlines the purpose of each phase of the quality assurance system. This outline is intended to show only the relationship of mix design testing to the overall program of job inspection and control. More discussion of quality management is contained in Chapter 8. The actual details of field inspection may be found in the Asphalt Institute's *Principles of Construction of Hot-Mix Asphalt Pavements,* Manual Series No. 22 (MS-22).

(a) *Pre-Production Sampling and Testing*—The principal purpose of this phase is to determine that the prospective sources of aggregate and asphalt are of satisfactory quality and will produce a paving mix satisfying all of the physical requirements and mix design requirements contained in the specifications.

(b) *Job-Mix Formula Verification*—In this phase, tests are performed at the start of plant production to compare field-produced mixture properties with the previously-established job-mix formula that was based on laboratory-mixed specimens. This is one of the key points of quality control for the paving construction since the job-mix formula establishes the actual gradation and asphalt content of the production mix. It may be necessary to make slight adjustments in the job-mix formula due to the characteristics of aggregate components. If the specifications are met, these results may then serve as the new adjusted job-mix formula and would then be the accepted target for all quality control testing that occurs.

> When a field laboratory is required by the contracting agency and used for the purpose of job mix control, it should meet the same requirements for test equipment and test procedures as a central laboratory.

(c) *Daily Job-Control Testing*—Quality control testing is performed during production to indicate if any of the mix properties deviate from the specifications. This testing is performed on an established schedule during the paving operation. Representative samples of the hot mix asphalt are obtained at the mixing facility and analyzed for design properties. (A random sampling method should be used to obtain HMA samples.) The results are compared with the job-mix control specifications. When irregularities occur and the limits of the job-mix formula are exceeded, appropriate corrections may be required at the plant. Occasionally, situations may warrant re-evaluation and redesign of the paving mixture.

(d) *In-Place Acceptance Testing*—Acceptance sampling and testing of in-place HMA can be authorized by the specifying agency to assure that satisfactory quality control has been exercised to attain the proper specification compliance. Its importance is emphasized by the fact that the results of these tests serve as a basis for the final acceptance of the paving construction by the owner.

2.07 AGGREGATE SIZE FRACTIONS — It is almost universal practice to specify the gradation of aggregates on the basis of the total aggregate gradation, i.e., total percent by weight passing the designated sieve sizes. The individual fractions of the total aggregate gradation, however, are typically designated in terms such as:

Coarse Aggregate [retained on 2.36 mm (No. 8) sieve]
Fine Aggregate [passing 2.36 mm (No. 8) sieve]
Mineral Filler [passing 75μm (No. 200) sieve]

It is also important to note that the aggregate gradations and individual fractions are specified independently of the total mix composition or binder content; i.e., the total aggregate equals 100 percent.

Aggregate materials often are also identified in broader terms such as rock, sand, and filler, or in terms of aggregate size designations as supplied by the aggregate producer, such as 57s, 68s, and 8s. These terms usually are applied to the stockpiled materials supplied to the job site. These definitions appear to have the greatest usage:

Rock—Material that is predominantly coarse aggregate [retained on 2.36 mm (No. 8)]
Sand—Material that is predominantly fine aggregate [passing 2.36 mm (No. 8)]
Filler—Material that is predominantly mineral dust [passing 75 μm (No. 200)]

Chapter 3 presents further information on aggregate gradations and blending calculations.

2.08 PROPORTIONING AGGREGATE AND ASPHALT — When proportioning asphalt and aggregate it is important to note that the asphalt content may be expressed either as a percentage by weight of total mix or as a percentage by weight of dry aggregate. While expressing asphalt content as a percentage by weight of total mix is most common, each method of specifying asphalt has certain advantages and either method is acceptable provided it is clearly understood which method is being used. This example illustrates the two methods:

Assume that a given mix contains aggregates and asphalt in the proportion of 45.36 kg (100 lb.) of aggregate to 2.72 kg (6 lb.) of asphalt. The asphalt content of such a mix could be expressed as $(2.72 \div 45.36) 100 = 6.0$ percent asphalt by weight of dry aggregate. This asphalt content may also be expressed as $[2.72 \div (45.36 + 2.72)] 100 = 5.7$ percent asphalt by weight of total mix.

2.09 TESTING METHODS — The material testing methods will normally be part of the contract and/or project specifications. These test methods are recommended if others are not specified. (Test methods shall be the latest revision of methods adopted by the American Association of State Highway and Transportation Officials or the American Society for Testing and Materials.)

	ASTM Designation	AASHTO Designation
(a) Asphalt Cement		
Penetration	D 5	T 49
Viscosity		
Absolute	D 2171	T 202
Kinematic	D 2170	T 201
Flash Point	D 92	T 48
Thin Film Oven Test	D 1754	T 179
Rolling Thin Film Oven Test	D 2872	T 240
Ductility	D 113	T 51
Solubility	D 2042	T 44
Specific Gravity	D 70	T 228
(b) Mineral Aggregates		
Los Angeles Abrasion	C 131 or C 535	T 96
Unit Weight	C 29	T 19
Sieve Analysis (Aggregates)	C 136	T 27
Sieve Analysis (Filler)	D 546	T 37
Specific Gravity (Coarse)	C 127	T 85
Specific Gravity (Fine)	C 128	T 84
Specific Gravity (Filler)	D 854 or C 188	T 100 or T 133
Sulfate Soundness	C 88	T 104
Sand Equivalent	D 2419	T 176
Particle Shape	D 4791	
(c) Hot Mix Asphalt Paving		
Asphalt Content (extraction)	D 2172	T 164
Asphalt Content (nuclear)	D 4125	T 287
Recovery of Asphalt	D 1856	T 170
Density and Voids Analysis	See Chapter 4	
Maximum Specific Gravity of Paving Mixtures	D 2041	T 209
Bulk Specific Gravity	D 1188 or D 2726	T 166

2.10 MATERIAL SAMPLING AND TESTING — Prior to mix design testing, ample *representative* samples of aggregates and asphalt should be obtained to accomplish the required number of tests. These material quantities are suggested:

Asphalt Cement .... 4 liters (1 gal.)
Coarse Aggregate (or Rock) .... 25 kilograms (50 lb.)
Fine Aggregate (or Sand) .... 25 kilograms (50 lb.)
Filler (when required) .... 10 kilograms (20 lb.)

Additional materials may be required if the above quantities result in appreciable waste when combining materials for the design gradation, or if several aggregate combinations are to be investigated, or if water sensitivity analysis of the asphalt mixture is to be performed.

Each material sample should be completely identified by source location, project location, and project number or job number. Each asphalt cement sample should be stored in clean, small metal containers with tight lids or covers to avoid the necessity of reheating the entire supply each time a mix is tested. Each aggregate sample should be placed in a tightly woven cloth sack, securely wired or tied.

In advance of the mix design testing, a list or schedule of the tests to be performed should be made. Tests should be performed in proper and logical sequence. It is important that all material specification tests be completed before the mix design tests are started. In this way sources of substandard aggregates are eliminated from the design studies. Chapter 3, Evaluation of Aggregate Gradation, suggests a schedule of aggregate analysis and testing.

2.11 PREPARATION OF TEST MIXES — Detailed procedures used in the preparation of test mixes for each method of mix design are outlined later in this manual. In general, the procedures illustrated in Figures 2.6 through 2.9 are used.

Figure 2.6 – **Weighing aggregates for batch mixes**

Figure 2.7 – **Heating aggregate batches in oven**

Figure 2.8 – **Adding asphalt to heated aggregate**

Figure 2.9 – **Mechanical mixer for batch mixing of asphalt and aggregate**

Evaluation of Aggregate Gradation

3.01 GENERAL — Aggregate gradation analysis and the combining of aggregates to obtain the desired gradation are important steps in hot mix asphalt design. The aggregate gradation must meet the gradation requirements of the project specifications and yield a mix design that meets the criteria of the mix design method. The gradation should also be made up of the most economical aggregates available that are of suitable quality. Quality of aggregate particles in terms of physical characteristics is discussed in Chapter 7.

This chapter outlines the recommended steps for analyzing aggregates for asphalt paving mix design. The methods illustrated by these examples are applicable to blending and adjusting the aggregate gradation in laboratory control of the mix, in production control of aggregates, and in plant control during construction.

3.02 ANALYZING AGGREGATE FOR PRELIMINARY MIX DESIGNS — For preliminary mix designs, aggregate analysis will be governed, to some extent, by the method of producing the gradation during construction. Different methods may be used, depending on the type of local aggregate sources and the project specifications. In smaller operations, aggregates from local sources are sometimes produced either as a "single aggregate," or as a "major aggregate" supplemented by minor additions of filler or "sized aggregates." "Single aggregate" production operations will often require a certain amount of "waste" to obtain the desired grading. In larger operations, aggregates from commercial producers are usually supplied as "sized" aggregates, in which case all of the various required sizes are blended, with or without filler, to produce the aggregate gradation desired.

The laboratory procedures below normally apply to the testing of aggregates for preliminary mix designs. It may be necessary to modify this plan to meet the testing requirements that arise during the progress of the mix design.

(a) Dry all aggregate samples to constant weight at 105°C to 110°C (220°F to 230°F). Separate pans should be used for each aggregate sample.

(b) Perform washed sieve analysis and specific gravity tests on representative samples including filler from each proposed aggregate source.

(c) Compute a blend of aggregates to produce the desired mix gradation, using the gradation for each aggregate source (adjusted for waste if required). A starting point for preliminary mix designs would be an aggregate gradation that approaches the median of the specification limits.

(d) Separate each dried aggregate into fractions (sizes) if necessary based on the results of (b). These sizes are generally recommended:

plus 19.0 mm (3/4 in.)
19.0 mm (3/4 in.) to 9.5 mm (3/8 in.)
9.5 mm (3/8 in.) to 4.75 mm (No. 4)
4.75 mm (No. 4) to 2.36 mm (No. 8)
minus 2.36 mm (No. 8)

The size separations may vary depending on the amount of material in each fraction for a specific aggregate. Combine fractions if necessary.

(e) Compute the blend proportions and batch weights (see Article 3.13) of the sized aggregates and filler required to produce batch mixes of the desired gradation. As a matter of practical convenience it is preferable to use the same weight of aggregate for each batch in the trial mixes.

(f) Prepare mix design test specimens in accordance with the procedure prescribed for the particular mix design method being used.

3.03 AGGREGATE ANALYSIS FOR JOB-MIX FORMULA — To determine the job-mix formula, the aggregate analysis will be somewhat governed by the number of aggregate stockpiles and the type of hot mix asphalt mixing facility being used. This phase of mix design establishes the job-mix formula that defines the actual gradation and asphalt content to be obtained in the finished construction.

The procedures below normally apply to the testing of aggregates for establishing the job-mix formula at the mixing facility. More discussion on verification of the asphalt mixture at the mixing facility is found in Chapter 8. For the actual details of field sampling, plant calibration and control, refer to the Asphalt Institute's *Principles of Construction of Hot Mix Asphalt Pavements* (MS-22). It may be necessary to modify this plan to suit the control features of the mixing facility.

(a) Secure representative samples from each aggregate stockpile, including filler, to be used in the production of the hot mix asphalt.

(b) Dry all aggregate samples to constant weight at 105°C to 110°C (220°F to 230°F). Separate pans should be used for each aggregate sample.

(c) Perform washed sieve analysis and specific gravity tests on representative samples from the respective stockpiles.

(d) Compute the blend of aggregates required to produce the desired mix gradation, using the full gradation for each individual aggregate.

(e) Adjust the cold aggregate feeder controls to obtain the desired aggregate blend and combined gradation.

(f) In a batch plant, perform a washed sieve analysis on a representative sample of each aggregate size separation produced. These samples should be obtained from the hot bins only after the gradation unit has reached normal operation. In a drum-mix plant, perform a washed sieve analysis on representative samples taken from the main cold-feed belt.

(g) Compute the blend proportions and batch weights of the sized aggregates (if a batch plant) or combined aggregate (if a drum plant) and filler required to produce one laboratory batch mix of the desired gradation. It is preferable to use the same weight of aggregate for each batch in a trial mix series.

(h) Prepare test specimens of the design mix in accordance with the procedure prescribed for the mix design method being used. **Analyze the test specimens to determine that the mixture has the same properties as that designed in the laboratory.**

(i) Adjust plant controls to obtain the design asphalt content and blend proportions of aggregates desired in the final paving mix.

(j) Verify the aggregate gradation in the plant mix by performing washed sieve analysis on representative samples of extracted aggregate.

3.04 BLENDING AGGREGATES BY WEIGHT — Determining the proportions of two or more aggregates to achieve a gradation within the specification limits is largely a matter of trial and error. Graphical and mathematical methods may sometimes be used to advantage. It is desirable to initially plot the sieve analyses for all aggregates to be used as shown in Figures 3.1 and 3.2. In this way it is often possible to make a visual estimate of the blend proportions required, depending on the experience with the local aggregates.

3.05 BASIC EQUATION — Regardless of the number of aggregates combined or of the method by which the proportions are determined, the basic equation expressing the combination is:

$$P = Aa + Bb + Cc + ...\qquad(1)$$

where, P = the percentage of the combined aggregates passing a given sieve;
$A, B, C, ...$ = percentage of material passing a given sieve for the individual aggregates; and
$a, b, c, ...$ = proportions of individual aggregates used in the combination, where the total = 1.00.

The combined percentages (P in Eq. 1) for each of the different sieve sizes should closely agree with the desired percentages for the combined aggregate. None should fall outside the established grading specification limits. Obviously, there may be several acceptable combinations. An optimum combination would obviously be one in which the percentages of the blend are in as close agreement as possible to the original desired percentages. However, the desired percentages are difficult to define and could change with specific project circumstances.

Mathematical procedures have been developed that will directly determine an "optimum" combination of aggregates. To accomplish this calculation, various optimization procedures have been used, such as the closest to the middle of the specification range. The speed of these methods varies with the number of stockpiles

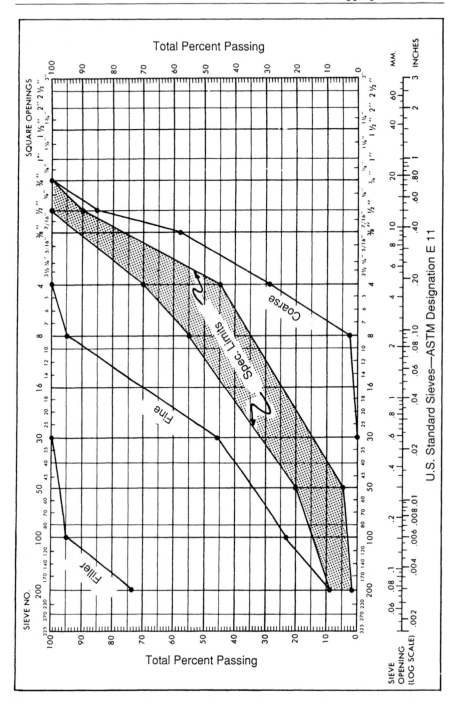

Figure 3.1 – **Job aggregates and specification plotted on conventional aggregate grading chart**

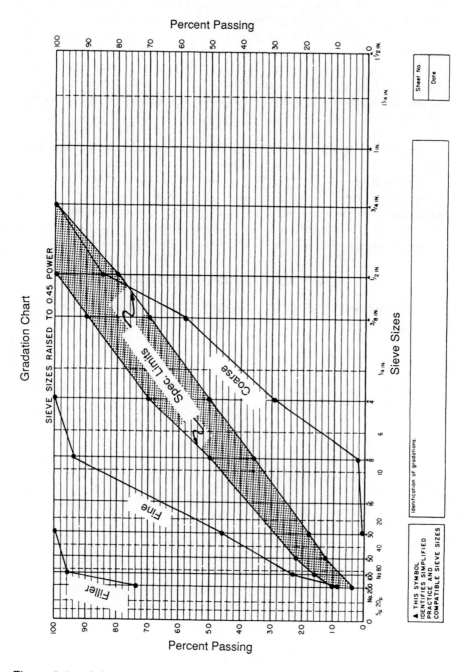

Figure 3.2 – **Job aggregates and specification plotted on 0.45 power gradation chart**

being blended. Many designers use computer spreadsheet programs to assist in quickly evaluating a number of alternative blends. The Asphalt Institute's *Computer-Assisted Asphalt Mix Analysis* (CAMA) computer program can also be used to visually evaluate the gradation plot of numerous blends very rapidly. Regardless of which method is used, a trial-and-error approach guided by a certain amount of reasoning is usually the easiest and best procedure to determine and refine a satisfactory combination of aggregates. A computer-generated blend should not be used without further evaluation and checking.

3.06 GRAPHICAL SOLUTIONS — Graphical methods have also been devised for determining combinations of aggregates to obtain a desired gradation. Like mathematical methods, some graphical methods are quite complicated. As the number of aggregates to be combined is increased, the graphical method becomes even more complicated. For two and sometimes three aggregate materials, graphical solutions may be used to advantage over trial-and-error methods. In other cases, graphical methods may be used to indicate the starting point for trial-and-error solutions.

Proportioning Determinations

3.07 COMBINING TWO AGGREGATES — The basic equation for combining two aggregates is:

$$P = Aa + Bb \tag{2}$$

Since a + b = 1, then a = 1 - b. Substituting this expression into Eq. 2 and solving for b:

$$b = \frac{P - A}{B - A} \tag{3}$$

An expression for a, can also be found:

$$a = \frac{P - B}{A - B} \tag{4}$$

EXAMPLE

Assume that a single aggregate stockpile is to be blended with sand to meet grading requirements for an asphalt paving mixture. These are given in Figure 3.3a as aggregates A and B. To make a determination:

1. Examine the two gradations to determine which aggregate will contribute most of the material for certain sizes. In this case, most of the minus 2.36 mm (No. 8) aggregate will be furnished by aggregate B.

2. Using the percentages for the 2.36 mm (No. 8) sieve and substituting into Eq. 3, the proportions are determined to meet the midpoint of the specification (Figure 3.3b).

3. Inspection of the first trial gradation shows the percent passing the 75 μ m (No. 200) sieve to be close to the lower specification limit. Increase the proportion of aggregate B (in this case to 0.55) and compute the gradation of the second trial (Figure 3.3c).

4. Inspection now shows the gradation is critical on the 600 μ m (No. 30) sieve. Reduce the proportion of aggregate B to 0.52 and compute the gradation of the third trial (Figure 3.3d). The third trial is a "best fit" considering the percents passing the 600 μ m (No. 30) and 75 μ m (No. 200) sieves.

The two aggregates may also be combined graphically, as shown in Figure 3.4:

1. The percents passing the various sizes for aggregate A are plotted on the right-hand vertical scale (representing 100 percent aggregate A).

2. The percents passing the various sizes for aggregate B are plotted on the left-hand vertical scale (representing 100 percent aggregate B).

3. Connect the points common to the same sieve size with straight lines, and label.

4. For a particular size, indicate on the straight line where the line crosses the specification limits measured on the vertical scale. (As an example, for the 9.5 mm (3/8-in.) size, two points, shown as arrows, are plotted on the line at 70 and 90 percent on the vertical scale.)

5. That portion of the line between the two points represents the proportions of aggregates B and A, indicated on the top and bottom horizontal scales, that will not exceed specification limits for that particular size.

6. The portion of the horizontal scale designated by two vertical lines, when projected within specification limits for all sizes, represents the limits of the proportions possible for satisfactory blends. In this case, 43 to 54 percent of aggregate A and 46 to 57 percent of aggregate B will meet specifications when blended. It can also be seen that the percent of blended material passing the 600 μ m (No.30) and 75 μ m (No. 200) sieves will be the critical or controlling sizes for keeping the blend within the specification limits.

7. For blending, usually the midpoint of that horizontal scale is selected for the blend. In this case, 48 percent aggregate A and 52 percent aggregate B.

PERCENT PASSING

Sieve	19.0 mm	12.5 mm	9.5 mm	4.75 mm	2.36 mm	600 μm	300 μm	150 μm	75 μm
Sieve	3/4"	1/2"	3/8"	No.4	8	30	50	100	200
Spec.	100	80-100	70-90	50-70	35-50	18-29	13-23	8-16	4-10
Aggr. A	100	90	59	16	3.2	1.1	0	0	0
Aggr. B	100	100	100	96	82	51	36	21	9.2

(a) Grading Specification and sieve analyses of aggregates

For 2.36mm (No.8), $b = \dfrac{P-A}{B-A} = \dfrac{42.5-3.2}{82-3.2} = 0.50, a = 1 - 0.50 = 0.50$

Sieve	19.0 mm	12.5 mm	9.5 mm	4.75 mm	2.36 mm	600 μm	300 μm	150 μm	75 μm
Sieve	3/4"	1/2"	3/8"	No.4	8	30	50	100	200
.50 x A	50	45.0	29.5	8.0	1.6	0.6			
.50 x B	50	50.0	50.0	48.0	41.0	25.0	18.0	10.5	4.6
Total	100	95.0	79.5	56.0	42.6	25.6	18.0	10.5	4.6
Spec.	100	80-100	70-90	50-70	35-50	18-29	13-23	8-16	4-10

Minus 75μm(No.200)low, increase b to 0.55, a to 0.45

(b) First trial combination

Sieve	19.0 mm	12.5 mm	9.5 mm	4.75 mm	2.36 mm	600 μm	300 μm	150 μm	75 μm
Sieve	3/4"	1/2"	3/8"	No.4	8	30	50	100	200
.45 x A	45	40.5	26.6	7.2	1.4	0.5			
.55 x B	55	55.0	55.0	52.8	45.1	28.0	19.8	11.5	5.1
Total	100	95.5	81.6	60.0	46.5	28.5	19.8	11.5	5.1
Spec.	100	80-100	70-90	50-70	35-50	18-29	13-23	8-16	4-10

Minus 600μm (No.30) high, let b = 0.52, a = 0.48

(c) Second trial combination

Sieve	19.0 mm	12.5 mm	9.5 mm	4.75 mm	2.36 mm	600 μm	300 μm	150 μm	75 μm
Sieve	3/4"	1/2"	3/8"	No.4	8	30	50	100	200
.48 x A	48	43.2	28.3	7.7	1.5	0.5	0	0	0
.52 x B	52	52	52	49.9	42.6	26.5	18.7	10.9	4.8
Total	100	95.2	80.3	57.6	44.1	27.0	18.7	10.9	4.8
Spec.	100	80-100	70-90	50-70	35-50	18-29	13-23	8-16	4-10

(d) Third trial combination

Figure 3.3 – **Trial-and-error calculations for combining two aggregates**

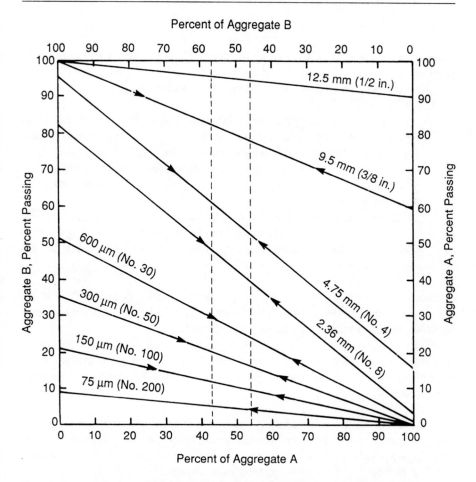

Figure 3.4 – **Graphical solution for proportioning two aggregates**

3.08 COMBINING THREE AGGREGATES — The basic equations for combining three aggregates are:

$$P = Aa + Bb + Cc \tag{5}$$

$$1.0 = a + b + c \tag{6}$$

The mathematical and graphical trial and error procedures are more complicated than with two aggregates, as expected.

EXAMPLE

Assume that mineral filler, C, is to be blended with aggregates A and B from the previous example to obtain a gradation meeting specification requirements. The specification and gradations are given in Figure 3.5 and the procedure is:

1. An inspection of the gradations indicates that there is a reasonably clean separation between the plus 2.36 mm (No. 8) sieve sizes and the minus 2.36 mm (No. 8) sizes. Aggregate A will furnish most of the plus 2.36 mm (No. 8) sizes.
2. Determine the approximate proportion of aggregate A required to obtain 42.5 percent passing the 2.36 mm (No. 8) sieve (the midpoint of specification range), using Eq. 4 (See Figure 3.5).
3. The percentages passing the 75 μ m (No. 200) sieve are examined next using Eqs. 5 and 6. Values are substituted from Eq. 6 into Eq. 5, assuming a is 0.50 from above. The remainder of the calculations are self-explanatory and are shown in Figure 3.5b.

Should the blended gradation exceed specification limits, that proportion in the blend apparently responsible should be altered, with the other aggregate proportions altered similarly to make up a total of 1.0 or 100 percent.

Trial-and-error solutions are exactly that; however, each trial reveals more information to reduce the range of possible solutions. An inspection of the gradations for indications to assist in establishing proportions narrows the number of solutions considerably.

Graphical methods may help in trial-and-error solutions; however, they may not in cases of aggregate gradations having overlapping grading curves. Several graphical methods are possible, but the procedure described here appears to be one of the more practical procedures. Each of the aggregates is divided into these gradings:

(a) percent material retained on 2.36 mm (No. 8) sieve,
(b) percent material passing 2.36 mm (No. 8) but retained on the 75 μ m (No. 200) sieve, and
(c) percent material passing 75 μ m (No. 200) sieve.

The specification limits are divided in a somewhat similar manner: (corresponding to a) allowable percentage limits retained on a 2.36 mm (No. 8) sieve, and (corresponding to c) allowable percentage limits passing 75 μ m (No. 200) sieve.

PERCENT PASSING

Sieve	19.0 mm	12.5 mm	9.5 mm	4.75 mm	2.36 mm	600 μm	300 μm	150 μm	75 μm
	3/4"	1/2"	3/8"	No.4	8	30	50	100	200
Spec.	100	80-100	70-90	50-70	35-50	18-29	13-23	8-16	4-10
Aggr. A	100	90	59	16	3.2	1.1			
Aggr. B	100	100	100	96	82	51	36	21	9.2
Aggr. C	100	100	100	100	100	100	98	93	82

(a) Grading specification and gradations

$2.36mm$ (No. 8), $\quad a = \dfrac{P-B}{A-B} = \dfrac{42.5-82}{3.2-82} = 0.50$

$75 \mu m$ (No. 200), $\quad P = Aa + Bb + Cc$

$\qquad 7 = 0\,(0.50) + 9.2b + 82c\,, \qquad b+c = 1-0.50 = 0.50$

$\qquad 7 = 9.2\,(0.50-c) + 82c$

$\qquad c = \dfrac{7-4.6}{82-9.2} = 0.03\,, \qquad b = 0.50-0.03 = 0.47$

Sieve	19.0 mm	12.5 mm	9.5 mm	4.75 mm	2.36 mm	600 μm	300 μm	150 μm	75 μm
	3/4"	1/2"	3/8"	No.4	8	30	50	100	200
.50 x A	50	45.0	29.5	8.0	1.6	0.6			
.47 x B	47	47.0	47.0	45.1	38.5	24.0	16.9	9.9	4.3
.03 x C	3	3.0	3.0	3.0	3.0	3.0	3.0	2.8	2.5
Total	100	95.0	79.5	56.1	43.1	27.6	19.9	12.7	6-8
Spec.	100	80-100	70-90	50-70	35-50	18-29	13-23	8-16	4-10

(b) First trial solution for combining three aggregates

Figure 3.5 – Trial-and-error solution for combining three aggregates

EXAMPLE

To combine the same three aggregates in a graphical solution, points representing each of the three aggregate gradings are plotted on a chart, as shown in Figure 3.6. Only the portion retained on the 2.36 mm (No. 8) sieve and the portion passing the 75 μ m (No. 200) sieve are used for each of the three aggregates to locate points on the chart. The point designated A represents the coarse aggregate grading, B is the sand or fine aggregate grading, and C is the mineral filler grading. Point S represents the middle of the specification grading band for material retained on the 2.36 mm (No. 8) sieve and passing the 75 μ m (No. 200) sieve. Lines are drawn between and beyond Points A and S and between Points B and C. Line AS is extended to Line BC to establish Point B'. The length of each line segment is determined by using the differences in percentages between terminal points. Then the percentage of each aggregate material needed for

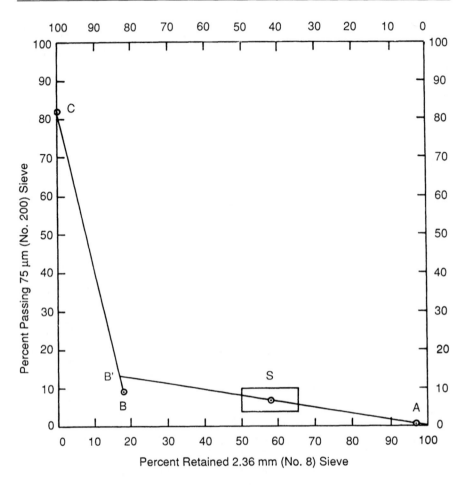

Figure 3.6 – **Chart for estimating three aggregate blends**

the blend is calculated using the following equations derived from the basic formula (Eq. 1).

$$a = \frac{\text{line SB'}}{\text{line AB'}} = \frac{58 - 77}{97 - 17} = \frac{41}{80} = 0.51$$

Since $b + c = 1.0 - a$, then

$$c = \frac{(1.0 - a)\ \text{line B'B}}{\text{line CB}} = \frac{(1.0 - 0.51) \times (13 - 9)}{82 - 9} = \frac{0.49 \times 4}{73} = 0.03$$

$$b = 1.0 - a - c = 1.0 - 0.51 - 0.03 = 0.46$$

3.09 ADJUSTING FOR SUFFICIENT VOID SPACE — A link between the aggregate gradation curve and asphalt mixture performance has been recognized for many years. However, quantifying this relationship has been difficult. Considerable research has been done to try to find a formula for the "optimum" gradation curve that would provide adequate space for a minimum amount of asphalt and air voids as well as adequate stability under traffic. Some sources have shown that this combined space for air and asphalt could not be used to statistically explain observed performance; this is due to the many other factors that can affect overall pavement performance. Regardless, the concept of needing sufficient space to develop adequately-thick asphalt films on the aggregate for adhesion and durability is fundamentally sound.

The packing characteristics of asphalt-coated aggregate particles in an asphalt mixture are related to both aggregate surface characteristics and gradation. Aggregate surface characteristics include angularity and surface texture. When selecting aggregate for a project, surface properties are not considered for void purposes, but rather for stability and skid resistance. Therefore, if adjustments in void space are required, changes are usually made to the aggregate gradation. Unfortunately, the guidance for adjusting the gradation curve to change the void space is not straightforward.

It has been shown that the densest packing condition of aggregate particles is approximated by the straight line on the 0.45 power chart as discussed in Chapter 2 and shown in Figure 2.5. A combined aggregate gradation that plots very close to this line is not desirable. These blends typically produce very low voids in the mineral aggregate (VMA) which is the space occupied by both the asphalt cement and the air voids. (See Chapter 4 for more information on VMA.) Even if aggregate blends near this line meet the aggregate grading specification, a gradation curve that closely follows the maximum density line should be avoided.

If too much void space is measured in a laboratory-compacted mix and an open-graded mixture was not being designed, then the aggregate blend could be adjusted to plot slightly closer to this line to possibly decrease the void content. In this approach, it is assumed that the problem lies with the gradation and not with the laboratory compaction procedure. Aggregate adjustments are discussed in Articles 3.07 and 3.08.

The more common problem is too little VMA, which can lead to inadequate space for the asphalt cement. Many approaches have been tried to increase VMA by adjusting the gradation.

If all other factors remain constant, fine aggregate contributes more to VMA than coarse aggregate. In practice, changes are often attempted by adding natural sands, which normally are fine, well-rounded particles passing the 600 μ m (No. 30) sieve. However, excessive amounts of natural sand have been identified as a cause of permanent deformation and tender mix problems, as discussed in Article 3.10. The addition of natural sand to increase VMA is strongly discouraged. Many agencies have placed maximum limits on the percentage of fine, natural sand, such as 15 to 20 percent of the total weight of the aggregate. These rounded particles, which include more inherent void space than manufactured, angular sands, may also allow compaction to occur more easily and thoroughly. This can lead to a decrease in VMA, defeating the purpose.

The filler material [particles passing the 75 µm (No.200) sieve] is the aggregate component with the highest VMA due to its large surface area. These VMA values have been reported to be as high as 32 percent. However, adding more of these fines to the mix can produce different VMA results, because of the wide variety of shapes and sizes found in these particles. In some cases, the extremely fine particles (less than 10 microns) may function as an asphalt extender which would effectively cause the available VMA to decrease, not increase as desired. Most specifications limit the amount of the filler- sized material; therefore, it is not usually feasible to increase VMA by increasing the percentage of these particles.

VMA increases can be achieved by an overall adjustment of the gradation or possibly changing the shape or texture of the intermediate portion of the blended aggregates. Because of the interaction of the two factors on the packing characteristics of an asphalt mixture, gradation changes (typically moving away from the maximum density line) are only reasonable with the same aggregate sources. By adjusting the proportional percentages of the aggregates that substantially contribute to the intermediate sizes, the gradation curve can be revised to plot further away from the maximum density line. Again, the previous examples with two and three aggregates demonstrate the mechanics of such an adjustment. In most cases, this shift will increase VMA.

Particle shape and texture can also make a difference. Changing the source of one aggregate may introduce a completely new interaction between all of the aggregate particles. Specifically, changing the angular shape and texture of coarse aggregates by crushing, or switching from natural sands to more angular manufactured sands (or screenings), can implement a significant change in VMA. The entire aggregate interlocking and packing mechanism is modified. In each specific locale, engineers and material suppliers should develop experience with the behavior of local materials. In summary, the same aggregate gradation with the same compaction effort, but with different shaped particles, can produce a different amount of VMA.

3.10 CHECKING FOR TENDER MIXES — A pavement that exhibits very low resistance to deformation under relatively heavy loads with small contact areas or that shoves and scuffs with tight radius, turning loads is showing the symptoms of a tender mix. This phenomenon will usually occur very early in the pavement service life. Fortunately, this problem can be temporary; it may disappear as the asphalt cement ages and the mix toughens with the kneading action of traffic.

There are many factors which may contribute to this behavior, as discussed in *Tender Mixes: The Causes and the Cures,* Asphalt Institute Information Series No. 168 (IS-168). Aggregate surface characteristics are again significant. Rounded aggregates are far more likely to have this problem than angular, rough-textured aggregates. Normally, there is a combination of factors involved. Since aggregate interparticle contact provides nearly all of the internal shear deformation of an asphalt mix, two of the most prevalent factors deal with the aggregate gradation.

One factor that is easily avoided is a low percentage of material passing the 75 µm (No. 200) sieve. Together with the asphalt cement, this portion of the aggregate makes a major contribution to the mix cohesion. High cohesion provides the internal tensile strength and mix toughness to resist the shearing forces. Both the size distribution and the percentage of these fines can have an additional impact on this effect.

The other factor that appears to be the most important single factor leading to mix tenderness is an excess of the middle-sized sand fraction in the material that passes the 4.75 mm (No. 4) sieve. This condition is characterized by a "hump" in the grading curve that can appear on nearly any sieve between the 4.75 mm (No. 4) and the 150 μ m (No. 100) sieve. The hump typically occurs near the 600 μ m (No. 30) sieve. Problem gradations can usually be detected on the 0.45 power gradation chart. Studies have indicated that if there is a deviation exceeding 3 percent upward in the gradation curve from a straight line drawn from the origin of the chart to the point at which the gradation curve crosses the 4.75 mm (No. 4) sieve line, a tender mix problem is likely. An example of this occurrence is shown in Figure 3.7.

This gradation check may not apply to highly-crushed, rough-textured aggregates. It should be noted that gradation alone may not cause tenderness in a mix. If other factors exist at the same time, this "humped" type of gradation can be a significant consideration. It is recommended that an aggregate gradation curve with this shape be avoided.

3.11 ADJUSTING FOR DIFFERENT SPECIFIC GRAVITIES — Aggregate gradations and grading curves are determined and expressed in percentages of total weight for convenience of measurement. The grading curve of each material is determined by weight using sieve analysis. However, grading specifications are established to require certain percentages of material at each of the various volumetric

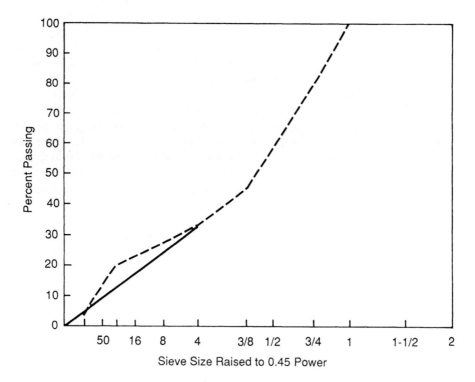

Figure 3.7 – **Example of a "humped" gradation**

Aggregate	Sp Gr	Proportion
A	1.00	0.52
B	2.00	0.45
C	3.00	0.03

Aggr.	Percent Vol.	Sp Gr	(1) Weight	(2) Percent Wt.
A	52.0	1.00	52.00	34.4
B	45.0	2.00	90.00	59.6
C	3.0	3.00	9.00	6.0
Total	100.0	---	151.00	100.0

(1) Weight   = Vol. × Sp Gr

(2) Percent Wt. = $\dfrac{\text{Individual Weight, W}}{\text{Total Weight}} \times 100 = \dfrac{W}{151} \times 100$

Figure 3.8 – **Adjusting percentages by volume to percentages by weight**

sizes in the hot mix asphalt. The blending process assumes that all of the aggregates have the same specific gravity. As long as the specific gravities of the combined aggregate materials are reasonably alike, the percentages by weight will approximate the percentages by volume for all practical purposes.

However, when the specific gravities of the individual aggregates differ significantly (by 0.20 or more) and they are to be blended together to make a target gradation, the aggregate proportions that are set up in the plant controls (percentages by weight) should be adjusted for this variance. Without this correction, the blend being made by weight at the plant to produce final volumetric proportions in the mix could deviate significantly from the blend of sizes designed in the laboratory on paper, since the different sized materials have different unit weights or densities. The adjustment is based on the fact that:

VOLUME X SPECIFIC GRAVITY = WEIGHT

EXAMPLE

Assume this combination was calculated for three aggregates having different specific gravities as shown in Figure 3.8. The necessary calculations are included in tabular form in the figure. The final percentages by weight in column 5 are the proportions to be used at the plant to obtain the percentages by volume in column 2 that were calculated by blending, assuming that all the specific gravities were the same.

3.12 ADJUSTING BY WASTING — Where the main source of aggregate is a single pit, it is often the case that the crusher-run aggregates are either coarser or finer than desired. If the gradation is coarser than desired, finer material can usually be blended with the crusher-run aggregate. But for gradations that have an excess of fines, the most economical adjustment is usually made by wasting a portion of the fine fraction. Most crushing plants will make the separations on the 4.75 mm (No. 4) [or possibly 2.36 mm (No. 8)] screen. Where an excess on a smaller size occurs, the correction is made by wasting a portion of the minus 4.75 mm (No. 4) [or 2.36 mm (No. 8)] fraction. The amount of waste is expressed as a percent, considering the total crusher-run material as 100 percent.

The equations for analysis of gradations before and after wasting are:

(Sizes above waste screen use percent retained)

$$R_b = \frac{R_2\,R_a}{R_1} \tag{7}$$

(Sizes below waste screen use percent passing)

$$P_b = \frac{P_2\,P_a}{P_1} \tag{8}$$

where, P_a, R_a = percent passing, or retained, of a given size before wasting;
P_b, R_b = adjusted percent passing, or retained, of a given size after wasting;
P_1, R_1 = percent passing, or retained, of the waste size before wasting; and
P_2, R_2 = adjusted percent passing, or retained, of the waste size after wasting.

The percent of waste, W, is found using:

$$W = \frac{100(P_1 - P_2)}{(100 - P_2)} \tag{9}$$

EXAMPLE

Assume that a single aggregate stockpile is being produced from a local roadside pit. The specification limits and crusher-run gradation are shown in Figure 3.9a. Note that the 4.75 mm (No. 4) size is above specification limits and that the other percentages approach the upper limits of the specification. A portion of the minus 4.75 mm (No. 4) fraction will be wasted to reduce the percent passing 4.75 mm (No. 4) from 75 to 70. The adjusted percentages of the sizes below the 4.75 mm (No. 4) sieve are found using Eq. 8 as shown. The other percentages are first converted to percent-retained and the adjusted percents retained on the sizes above the 4.75 mm (No. 4) are determined. These percentages of the coarse sizes are found using Eq. 7 and reconverting to percent passing. The percent of waste of the passing 4.75 mm (No. 4) fraction is then found using Eq. 9.

	19.0 mm	12.5 mm	9.5 mm	4.75 mm	2.36 mm	600 μm	150 μm	75 μm
Sieve	3/4"	1/2"	3/8"	#4	8	30	100	200
Spec.	100	80-100	70-90	55-73	40-55	20-30	10-18	4-10
% Pass, Pa	100	98	87	75	54	28	17	9
% Ret, Ra	0	2	13	25				
Adj % Ret, Rb	0	2	16	30				
Adj % Pass, Pb	100	98	84	70	50	26	16	8.4

$$P_b = \frac{P_2}{P_1} P_a = \frac{70}{75} P_a = 0.93 P_a \qquad R_b = \frac{R_2}{R_1} R_a = \frac{30}{25} R_a = 1.20 R_a$$

$$Waste, \ W = \frac{100(P_1 - P_2)}{(100 - P_2)} = \frac{100(75-70)}{(100-70)} = 16.7\%$$

(a) Calculations based on adjusting percent passing 4.75mm (No.4)

	19.0 mm	12.5 mm	9.5 mm	4.75 mm	2.36 mm	600 μm	150 μm	75 μm
Sieve	3/4"	1/2"	3/8"	#4	8	30	100	200
Spec.	100	80-100	70-90	55-73	40-55	20-30	10-18	4-10
% Pass, Pa	100	95	85	70	53	31	16	9
% Ret, Ra	0	5	15	30				
Adj % Ret, Rb	0	6	18	37				
Adj % Pass, Rb	100	94	82	63	48	28	14	8.1

$$P_b = \frac{P_2}{P_1} P_a = \frac{28}{31} P_a = 0.90 P_a \qquad R_b = \frac{R_2}{R_1} R_a = \frac{37}{30} R_a = 1.23 R_a$$

$$Waste, \ W = \frac{100(P_1 - P_2)}{100 - P_2} = \frac{100(70-63)}{100-63} = 18.9\%$$

(b) Calculations based on adjusting percent passing 600μm (No.30)

Figure 3.9 – **Adjusting gradation by wasting**

EXAMPLE

Assume, in this case, that the 600 μ m (No. 30) size exceeds the specification limits. Therefore, a sufficient amount of the minus 4.75 mm (No. 4) fraction will be wasted to reduce the 600 μ m (No. 30) material from 31 percent to 28 percent (shown in Figure 3.9b). In this case, P_1 and P_2 for the fine fraction sizes are the values for the 600 μ m (No. 30) sieve. When the adjusted percentages are obtained, R_1 and R_2 for the 4.75 mm (No. 4) sieve are used for the coarse fraction material.

3.13 COMPUTING LABORATORY BATCH WEIGHTS — In the analysis of aggregates for a given mix design, the final operation is the computation of the laboratory batch weights. It is convenient to use the same weight of aggregate in each batch of a trial mix series; in this way, the only variable is the amount of asphalt cement to be added. There are many approaches to preparing the laboratory batches of aggregate; most laboratories determine the simplest and best procedure for the kinds of materials that they typically use.

The goal of batching is to closely match the laboratory aggregate blend to the final field aggregate blend. Tight control over the stockpile sampling and laboratory blending procedures will help achieve the close match. The degree of control required in the laboratory will depend on how narrowly the aggregate stockpiles are sized.

The four aggregate sample gradations in Figure 3.10 are representative of field stockpiles for a mix design. The first step is to determine the blend of these gradations to achieve the desired gradation. For this example, the percentages are given below the stockpile gradations. After the blend is determined, there are typically three ways that the aggregates can be combined to create each batch. The three methods will be given in decreasing order of control over the batching process.

The most control is exerted when each stockpile is separated into fractions. Typically, these five fractions are used:

plus 19.0 mm (3/4 in.)
19.0 mm (3/4 in.) to 9.5 mm (3/8 in.)
9.5 mm (3/8 in.) to 4.75 mm (No. 4)
4.75 mm (No. 4) to 2.36 mm (No. 8)
minus 2.36 mm (No. 8)

The weight of each stockpile fraction needed to make a batch is determined by multiplying the total batch weight (typically 1,200 g), times the percentage of each stockpile needed in the blend, times the percentage of each stockpile fraction. The percentage of each stockpile fraction and its batch weight is shown in Figure 3.10. If a fraction represents less than one percent of a stockpile and another stockpile is the same aggregate type, substitutions should not sacrifice any batching control. In this example, if the 57/68 and #8 aggregates are from the same source, the 4.75 mm to 2.36 mm (No. 4 to No. 8) fraction for the 57/68 stockpile could be taken from that portion of the #8 stockpile.

The next method does not separate the stockpiles into fractions, but just combines the total sample weight of each stockpile into the batch. If the stockpiles are mostly one size material, the degree of control lost in this batching process should not be too great. As with sampling the stockpile in the field, care must be made to ensure that a representative portion is incorporated into each sample batch.

The method with the potential for the least control entails proportionately mixing a large amount of each stockpile to create the desired blend, and then taking 1,200 g of this blend to make each batch. As previously, the same care must be taken to ensure a representative sample is used in each batch. An additional drawback with this method is that if the desired gradation is modified later, these combined materials cannot be reused in the mix design.

Regardless of the aggregate batching method used, the weight of asphalt cement required for each batch is determined by dividing the aggregate batch weight (i.e., 1,200 g) by one minus the asphalt content (expressed as a decimal). (See Figure 3.10).

		Percent Passing				
Sieve Size		Aggregate Stockpiles				
mm (in.)	57/68	#8	#10	Nat. Sand	Blend	Target
25.4 (1.0)	100	100	100	100	100	100.0
19.0 (3/4)	95.3	100	100	100	99.1	95.0
12.7 (1/2)	35.3	99.5	100	100	86.8	78.0
9.5 (3/8)	8.0	88.6	100	100	75.9	68.0
4.75 (No. 4)	1.5	18.0	97.0	100	38.7	50.0
2.36 (No. 8)	1.3	5.0	70.0	89.8	25.7	36.0
1.18 (No. 16)	1.3	3.5	46.0	72.5	18.5	25.0
0.6 (No. 30)	1.2	3.2	30.0	51.5	13.0	17.0
0.3 (No. 50)	1.2	3.1	22.0	18.3	8.0	12.0
0.15 (No. 100)	1.1	3.0	17.0	2.3	5.4	8.0
0.075 (No. 200)	1.1	2.8	14.9	1.2	4.7	5.0
Percentage of Total Blend	20.0	50.0	20.0	10.0	100	

	Stockpile Passing/Retained, %			
Fraction, mm (in.)	57/68	#8	#10	Nat. Sand
plus 19.0 (3/4)	4.7	0	0	0
19.0 to 9.5 (3/4 to 3/8)	87.3	11.4	0	0
9.5 to 4.75 (3/8 to No. 4)	6.5	70.6	3.0	0
4.75 to 2.36 (No. 4 to No. 8)	0.2	13.0	27.0	10.2
minus 2.38 (No. 8)	1.3	5.0	70.0	89.8
Total	100	100	100	100

	Stockpile Batch Weights, g			
Fraction, mm (in.)	57/68	#8	#10	Nat. Sand
plus 19.0 (3/4)	11.3	0	0	0
19.0 to 9.5 (3/4 to 3/8)	209.5	68.4	0	0
9.5 to 4.75 (3/8 to No. 4)	15.6	423.6	7.2	0
4.75 to 2.36 (No. 4 to No. 8)	0.5	78.0	64.8	12.2
minus 2.38 (No. 8)	3.1	30.0	168.0	107.8
Total	240	600	240	120

Asphalt Cement Batch Weights		
AC, pct by wt of total mix	AC, g	Total Batch wt, g
3.5	43.5	1243.5
4.0	50.0	1250.0
4.5	56.5	1256.5
5.0	63.2	1263.2
5.5	69.8	1269.8

Figure 3.10 – **Worksheet for computing laboratory batch weights**

Chapter 4

Volumetric Properties of Compacted Paving Mixtures

4.01 GENERAL — The volumetric properties of a compacted paving mixture [air voids (V_a), voids in the mineral aggregate (VMA), voids filled with asphalt (VFA), and effective asphalt content (P_{be})] provide some indication of the mixture's probable pavement service performance. As noted in Article 2.04, the intent of laboratory compaction is to *simulate the in-place density of HMA after it has endured several years of traffic.* How well the laboratory compaction procedure simulates either the compacted state immediately after construction or after years of service can be determined by comparing the properties of an undisturbed sample removed from a pavement with the properties of a sample of the same paving mixture compacted in the laboratory.

It is necessary to understand the definitions and analytical procedures described in this chapter to be able to make informed decisions concerning the selection of the design asphalt mixture. The information here applies to both paving mixtures that have been compacted in the laboratory, and to undisturbed samples that have been removed from a pavement in the field.

A comparison of field and laboratory compacted mix properties has been made in several research studies. Statistical analyses of these data have failed to establish one laboratory compaction method that consistently produces the closest simulation to the field for all of the measured properties. However, there is a trend toward the gyratory method of compaction based on these findings and other subjective factors. This is a very complicated issue. Compaction method, level of compaction, structural concerns, construction conditions and other influences can all make a difference in these comparisons. Assuming that a reasonable degree of simulation is achieved by whatever compaction procedures are used, it is universally agreed that the air void analysis is an important part of mix design.

4.02 DEFINITIONS — Mineral aggregate is porous and can absorb water and asphalt to a variable degree. Furthermore, the ratio of water to asphalt absorption varies with each aggregate. The three methods of measuring aggregate specific gravity take these variations into consideration. The methods are ASTM bulk, ASTM apparent, and effective specific gravities. The differences among the specific gravities come from different definitions of aggregate volume.

Bulk Specific Gravity, G_{sb} - the ratio of the weight in air of a unit volume of a permeable material (including both permeable and impermeable voids normal to the material) at a stated temperature to the weight in air of equal density of an equal volume of gas-free distilled water at a stated temperature. See Figure 4.1.

Apparent Specific Gravity, G_{sa} - the ratio of the weight in air of a unit volume of an impermeable material at a stated temperature to the weight in air of equal density of an equal volume of gas-free distilled water at a stated temperature. See Figure 4.1.

Effective Specific Gravity, G_{se} - the ratio of the weight in air of a unit volume of a permeable material (excluding voids permeable to asphalt) at a stated temperature to the weight in air of equal density of an equal volume of gas-free distilled water at a stated temperature. See Figure 4.1.

NOTE: The accuracy of specific gravity measurements for mix design is important.

Unless specific gravities are determined to four significant figures (three decimal places) an error in air voids value of as much as 0.8 percent can occur. Therefore, the Asphalt Institute recommends the use of weigh scales whose sensitivity will allow a mix batch weighing 1000 to 5000 grams to be measured to an accuracy of 0.1 gram.

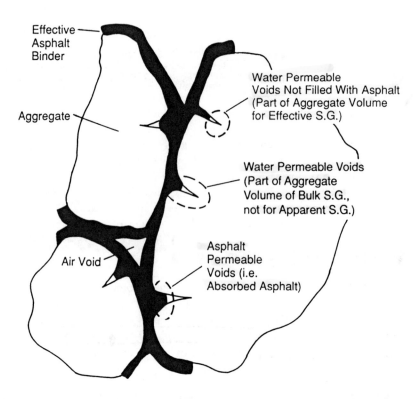

Effective Asphalt Binder

Water Permeable Voids Not Filled With Asphalt (Part of Aggregate Volume for Effective S.G.)

Aggregate

Water Permeable Voids (Part of Aggregate Volume of Bulk S.G., not for Apparent S.G.)

Air Void

Asphalt Permeable Voids (i.e. Absorbed Asphalt)

Figure 4.1 – **Illustrating bulk, effective, and apparent specific gravities; air voids; and effective asphalt content in compacted asphalt paving mixture**

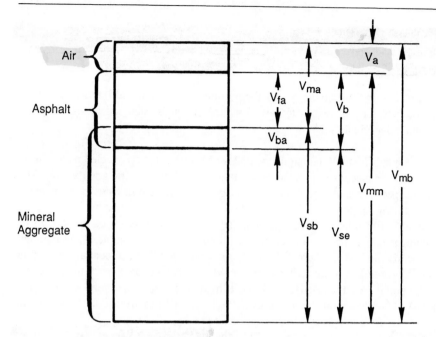

V_{ma} = Volume of voids in mineral aggregate
V_{mb} = Bulk volume of compacted mix
V_{mm} = Voidless volume of paving mix
V_{fa} = Volume of voids filled with asphalt
V_a = Volume of air voids
V_b = Volume of asphalt
V_{ba} = Volume of absorbed asphalt
V_{sb} = Volume of mineral aggregate (by bulk specific gravity)
V_{se} = Volume of mineral aggregate (by effective specific gravity)

Figure 4.2 – **Representation of volumes in a compacted asphalt specimen**

The definitions for voids in the mineral aggregate (VMA), effective asphalt content (P_{be}), air voids (V_a), and voids filled with asphalt (VFA) are:

Voids in the Mineral Aggregate, VMA - the volume of intergranular void space between the aggregate particles of a compacted paving mixture that includes the air voids and the effective asphalt content, expressed as a percent of the total volume of the sample. See Figure 4.2.

Effective Asphalt Content, P_{be} - the total asphalt content of a paving mixture minus the portion of asphalt that is lost by absorption into the aggregate particles. See Figure 4.2.

Air Voids, V_a - the total volume of the small pockets of air between the coated aggregate particles throughout a compacted paving mixture, expressed as percent of the bulk volume of the compacted paving mixture. See Figure 4.2.

Voids Filled with Asphalt, VFA - the portion of the volume of intergranular void space between the aggregate particles (VMA) that is occupied by the effective asphalt. See Figure 4.2.

The Asphalt Institute recommends that VMA values for compacted paving mixtures should be calculated in terms of the aggregate's bulk specific gravity, G_{sb}. The effective specific gravity should be the basis for calculating the air voids in a compacted asphalt paving mixture.

Table 4.1 illustrates that the type of aggregate specific gravity used in the analysis of a compacted paving mixture can have a very dramatic effect on the values reported for air voids and VMA. These differences are enough to make it appear that a mixture may satisfy or fail the design criteria for air voids and VMA depending on the aggregate specific gravity used for analysis. Asphalt Institute mix design criteria do not apply unless VMA calculations are made using bulk specific gravity and air void content calculations are made using effective specific gravity.

Voids in the mineral aggregate (VMA) and air voids (V_a) are expressed as percent by volume of the paving mixture. Voids filled with asphalt (VFA) is the percentage of VMA that is filled by the effective asphalt. Depending on how asphalt content is specified, the effective asphalt content may be expressed either as percent by weight of the total weight of the paving mixture, or as percent by weight of the aggregate in the paving mixture.

Because air voids, VMA and VFA are volume quantities and therefore cannot be weighed, a paving mixture must first be designed or analyzed on a volume basis. For design purposes, this volume approach can easily be changed over to a weight basis to provide a job-mix formula.

4.03 OUTLINE OF PROCEDURE FOR ANALYZING A COMPACTED PAVING MIXTURE — This list delineates all the measurements and calculations needed for a voids analysis:

   (a) Measure the bulk specific gravities of the coarse aggregate (AASHTO T 85 or ASTM C 127) and of the fine aggregate (AASHTO T 84 or ASTM C 128).

   (b) Measure the specific gravity of the asphalt cement (AASHTO T 228 or ASTM D 70) and of the mineral filler (AASHTO T 100 or ASTM D 854).

   (c) Calculate the bulk specific gravity of the aggregate combination in the paving mixture.

   (d) Measure the maximum specific gravity of the loose paving mixture (ASTM D 2041).

   (e) Measure the bulk specific gravity of the compacted paving mixture (ASTM D 1188 or ASTM D 2726).

   (f) Calculate the effective specific gravity of the aggregate.

   (g) Calculate the maximum specific gravity of the mix at other asphalt contents.

   (h) Calculate the asphalt absorption of the aggregate.

Table 4.1 – Influence of type of specific gravity on determination of VMA, VFA, and air voids

Bulk Specific Gravity of Compacted Mixture, G_{mb} 2.436
Density of Compacted Mixture, W_r, Mg/m^3 (lb/ft^3) 2.435 (152.0)
Asphalt Content, percent by weight of total mix 5.9
Asphalt Absorbed by Aggregate Particles, percent 0.8
Specific Gravity of Asphalt Cement, G_b 1.011

Specific Gravity Employed for Aggregate		Allowance For Absorption of Asphalt by Aggregate	Void Properties Compacted Mixture		
			Percent Voids in Mineral Aggregate	Percent Air Voids	Percent Voids Filled With Asphalt
ASTM Bulk	2.651	Yes	13.6	1.1	92
ASTM Bulk	2.651	No	13.6	-0.8	106
ASTM Bulk (sat. surf. dry)	2.716	Yes	15.6	3.2	79
ASTM Bulk (sat. surf. dry)	2.716	No	15.6	1.3	92
ASTM Apparent	2.834	No	19.1	4.9	74
Effective	2.708	No	15.4	1.1	93

(i) Calculate the effective asphalt content of the paving mixture.

(j) Calculate the percent voids in the mineral aggregate in the compacted paving mixture.

(k) Calculate the percent air voids in the compacted paving mixture.

(l) Calculate the percent voids filled with asphalt.

Equations for these calculations are found in Articles 4.05 through 4.11 and their application may be expedited by use of the appropriate worksheet, Figures 4.3 and 4.4.

4.04 PAVING MIXTURE DATA FOR SAMPLE CALCULATIONS — Table 4.2 illustrates the basic data for a sample of paving mixture. These design data are used in the sample calculations used in the remainder of this chapter.

4.05 BULK SPECIFIC GRAVITY OF AGGREGATE — When the total aggregate consists of separate fractions of coarse aggregate, fine aggregate, and mineral filler, all having different specific gravities, the bulk specific gravity for the total aggregate is calculated using:

$$G_{sb} = \frac{P_1 + P_2 + \ldots + P_n}{\dfrac{P_1}{G_1} + \dfrac{P_2}{G_2} + \ldots + \dfrac{P_n}{G_n}} \tag{1}$$

where, G_{sb} = bulk specific gravity for the total aggregate
P_1, P_2, P_n = individual percentages by weight of aggregate
G_1, G_2, G_n = individual bulk specific gravities of aggregate

Table 4.2 – **Basic data for sample of paving mixture**

(a) Constituents

Material	Specific Gravity		AASHTO Method	ASTM Method	Mix Composition	
	Bulk				Percent By Weight of Total Mix	Percent By Weight of Total Aggregate
Asphalt Cement	$1.030(G_b)$		T 228	D 70	$5.3(P_b)$	$5.6(P_b)$
Coarse Aggregate		$2.716(G_1)$	T 85	C 127	$47.4(P_1)$	$50.0(P_1)$
Fine Aggregate		$2.689(G_2)$	T 84	C 128	$47.3(P_2)$	$50.0(P_2)$
Mineral Filler	---		T 100	D 854	---	---

(b) Paving Mixture

Bulk specific gravity of compacted paving mixture sample, G_{mb}
(ASTM D 2726)   ————   2.442
Maximum specific gravity of paving mixture sample, G_{mm}
(ASTM D 2041)   ————   2.535

The bulk specific gravity of mineral filler is difficult to determine accurately. However, if the apparent specific gravity of the filler is substituted, the error is usually negligible.

Using the data in Table 4.2:

$$G_{sb} = \frac{50.0 + 50.0}{\frac{50.0}{2.716} + \frac{50.0}{2.689}} = \frac{100}{18.41 + 18.59} = 2.703$$

4.06 EFFECTIVE SPECIFIC GRAVITY OF AGGREGATE — When based on the maximum specific gravity of a paving mixture, G_{mm}, as measured using ASTM D 2041, the effective specific gravity of the aggregate, G_{se}, includes all void spaces in the aggregate particles except those that absorb asphalt. G_{se} is determined using:

$$G_{se} = \frac{P_{mm} - P_b}{\frac{P_{mm}}{G_{mm}} - \frac{P_b}{G_b}} \qquad (2)$$

where, G_{se} = effective specific gravity of aggregate
         G_{mm} = maximum specific gravity (ASTM D 2041) of paving mixture (no air voids)
         P_{mm} = percent by weight of total loose mixture = 100
         P_b = asphalt content at which ASTM D 2041 test was performed, percent by total weight of mixture
         G_b = specific gravity of asphalt

Using the data in Table 4.2:

$$G_{se} = \frac{100 - 5.3}{\dfrac{100}{2.535} - \dfrac{5.3}{1.030}} = \frac{94.7}{39.45 - 5.15} = 2.761$$

> NOTE: The volume of asphalt binder absorbed by an aggregate is almost invariably less than the volume of water absorbed. Consequently, the value for the effective specific gravity of an aggregate should be between its bulk and apparent specific gravities. When the effective specific gravity falls outside these limits, its value must be assumed to be incorrect. The calculations, the maximum specific gravity of the total mix by ASTM D 2041, and the composition of the mix in terms of aggregate and total asphalt content should then be rechecked for the source of the error.
>
> The apparent specific gravity, G_{sa}, of the total aggregate can be calculated by the same formula as the bulk by using the apparent specific gravity of each aggregate constituent.

4.07 MAXIMUM SPECIFIC GRAVITY OF MIXTURES WITH DIFFERENT ASPHALT CONTENTS — In designing a paving mixture with a given aggregate, the maximum specific gravity, G_{mm}, at each asphalt content is needed to calculate the percentage of air voids for each asphalt content. While the maximum specific gravity can be determined for each asphalt content by ASTM D 2041, the precision of the test is best when the mixture is close to the design asphalt content. Also, it is preferable to measure the maximum specific gravity in duplicate or triplicate.

After calculating the effective specific gravity of the aggregate from each measured maximum specific gravity (see Article 4.06) and averaging the G_{se} results, the maximum specific gravity for any other asphalt content can be obtained as shown below. For all practical purposes, the effective specific gravity of the aggregate is constant because the asphalt absorption does not vary appreciably with variations in asphalt content.

$$G_{mm} = \frac{P_{mm}}{\dfrac{P_s}{G_{se}} + \dfrac{P_b}{G_b}} \qquad (3)$$

where, G_{mm} = maximum specific gravity of paving mixture (no air voids)
P_{mm} = percent by weight of total loose mixture = 100
P_s = aggregate content, percent by total weight of mixture
P_b = asphalt content, percent by total weight of mixture
G_{se} = effective specific gravity of aggregate
G_b = specific gravity of asphalt

Using the specific gravity data from Table 4.2 and the effective specific gravity, G_{se}, determined in Article 4.06, the G_{mm} at an asphalt content, P_b, of 4.0 percent would be:

$$G_{mm} = \frac{100}{\dfrac{96}{2.761} + \dfrac{4.0}{1.030}} = \frac{100}{34.77 + 3.88} = 2.587$$

4.08 ASPHALT ABSORPTION — Absorption is expressed as a percentage by weight of aggregate rather than as a percentage by total weight of mixture. Asphalt absorption, P_{ba}, is determined using:

$$P_{ba} = 100 \frac{G_{se} - G_{sb}}{G_{sb}\, G_{se}} G_b \qquad (4)$$

where, P_{ba} = absorbed asphalt, percent by weight of aggregate
$\ G_{se}$ = effective specific gravity of aggregate
$\ G_{sb}$ = bulk specific gravity of aggregate
$\ G_b$ = specific gravity of asphalt

Using the bulk and effective aggregate specific gravities determined in Articles 4.05 and 4.06 and the asphalt specific gravity from Table 4.2:

$$P_{ba} = 100 \left(\frac{2.761 - 2.703}{2.703 \times 2.761}\right) 1.030 = 100 \left(\frac{0.058}{7.463}\right) 1.030 = 0.8$$

4.09 EFFECTIVE ASPHALT CONTENT OF A PAVING MIXTURE — The effective asphalt content, P_{be}, of a paving mixture is the total asphalt content minus the quantity of asphalt lost by absorption into the aggregate particles. It is the portion of the total asphalt content that remains as a coating on the outside of the aggregate particles and it is the asphalt content which governs the performance of an asphalt paving mixture. The formula is:

$$P_{be} = P_b - \frac{P_{ba}}{100} P_s \qquad (5)$$

where, P_{be} = effective asphalt content, percent by total weight of mixture
$\ P_b$ = asphalt content, percent by total weight of mixture
$\ P_{ba}$ = absorbed asphalt, percent by weight of aggregate
$\ P_s$ = aggregate content, percent by total weight of mixture

Using the data from Table 4.2 and Article 4.08:

$$P_{be} = 5.3 - \frac{0.8}{100} \times 94.7 = 4.5$$

4.10 PERCENT VMA IN COMPACTED PAVING MIXTURE — The voids in the mineral aggregate, VMA, are defined as the intergranular void space between the aggregate particles in a compacted paving mixture that includes the air voids and the effective asphalt content, expressed as a percent of the total volume. The VMA is calculated on the basis of the bulk specific gravity of the aggregate and is expressed as a percentage of the bulk volume of the compacted paving mixture. Therefore, the VMA can be calculated by subtracting the volume of the aggregate determined by its bulk specific gravity from the bulk volume of the compacted paving mixture. A method of calculation is illustrated for each type of mixture percentage content.

If the mix composition is determined as percent <u>by weight of total mixture:</u>

$$VMA = 100 - \frac{G_{mb} P_s}{G_{sb}} \tag{6}$$

where, VMA = voids in mineral aggregate, percent of bulk volume
G_{sb} = bulk specific gravity of total aggregate
G_{mb} = bulk specific gravity of compacted mixture
(AASHTO T166; ASTM D 1188 or D 2726)
P_s = aggregate content, percent by total weight of mixture

Using the data from Table 4.2 and Article 4.05:

$$VMA = 100 - \frac{2.442 \times 94.7}{2.703} = 100 - 85.6 = 14.4$$

Or if the mix composition is determined as percent <u>by weight of aggregate:</u>

$$VMA = 100 - \frac{G_{mb}}{G_{sb}} \times \frac{100}{100 + P_b} 100 \tag{7}$$

where, P_b = asphalt content, percent by weight of aggregate.

Using the data from Table 4.2 and Article 4.05:

$$VMA = 100 - \frac{2.442}{2.703} \times \frac{100}{100 + 5.6} \times 100 = 100 - 85.6 = 14.4$$

4.11 PERCENT AIR VOIDS IN COMPACTED MIXTURE — The air voids, V_a, in the total compacted paving mixture consist of the small air spaces between the coated aggregate particles. The volume percentage of air voids in a compacted mixture can be determined using:

$$V_a = 100 \times \frac{G_{mm} - G_{mb}}{G_{mm}} \tag{8}$$

where, V_a = air voids in compacted mixture, percent of total volume
       G_{mm} = maximum specific gravity of paving mixture (as determined in Article 4.07 or as measured directly for a paving mixture by ASTM D 2041)
       G_{mb} = bulk specific gravity of compacted mixture

Using the data from Table 4.2:

$$V_a = 100 \times \frac{2.535 - 2.442}{2.535} = 3.7$$

4.12 PERCENT VFA IN COMPACTED MIXTURE — The voids filled with asphalt, VFA, is the percentage of the intergranular void space between the aggregate particles (VMA) that are filled with asphalt. VFA, not including the absorbed asphalt, is determined using:

$$VFA = \frac{100\,(VMA - V_a)}{VMA} \tag{9}$$

where, VFA = voids filled with asphalt, percent of VMA
       VMA = voids in mineral aggregate, percent of bulk volume
       V_a = air voids in compacted mixture, percent of total volume

Using the data from Table 4.2 and Articles 4.10 and 4.11:

$$VFA = 100 \times \frac{14.4 - 3.7}{14.4} = 74.3 \text{ percent}$$

Figure 4.3 – **Worksheet: Analysis by weight of total mixture**

Worksheet for Volumetric Analysis of Compacted Paving Mixture
(Analysis by weight of total mixture)

Sample: _____   Date: _____

Identification: _____

Composition of Paving Mixture

		Specific Gravity, G		Mix Composition, % by wt. of Total Mix, P					
			Bulk		Mix or Trial Number				
					1	2	3	4	5
1. Coarse Aggregate	G_1		2.716	P_1			47.4		
2. Fine Aggregate	G_2		2.689	P_2			47.3		
3. Mineral Filler	G_3		--	P_3			--		
4. Total Aggregate	G_s	--	--	P_s			94.7		
5. Asphalt Cement	G_b	1.030	--	P_b			5.3		

	Equation*		
6. Bulk Sp. Gr. (G_{sb}), total aggregate	(1)		2.703
7. Max. Sp. Gr. (G_{mm}), paving mix ASTM D2041	—		2.535
8. Bulk Sp. Gr. (G_{mb}), compacted mix ASTM D2726	—		2.442
9. Effective Sp. Gr. (G_{se}), total aggregate	(2)		2.761
10. Absorbed Asphalt (P_{ba}), % by wgt. total agg.	(4)		0.8

CALCULATIONS

11. Effective Asphalt Content (P_{be}) =

$$\text{Line 5 } P_b - \frac{(\text{Line 10} \times \text{Line 4 } P_s)}{100}$$ (5)   4.5

12. VMA =

$$100 - \frac{\text{Line 8} \times \text{Line 4 } P_s}{\text{Line 6}}$$ (6)   14.4

13. Air Voids (V_a) =

$$100 \; \frac{\text{Line 7} - \text{Line 8}}{\text{Line 7}}$$ (8)   3.7

14. VFA =

$$100 \; \frac{\text{Line 12} - \text{Line 13}}{\text{Line 12}}$$ (9)   74.3

*Equations from Chapter 4, MS-2

Figure 4.4 – **Worksheet: Analysis by weight of aggregate**

Worksheet for Volumetric Analysis of Compacted Paving Mixture
(Analysis by weight of aggregate)

Sample: _____   Date: _____

Identification: _____

Composition of Paving Mixture

		Specific Gravity, G		Mix Composition, % by wt. of Aggregate, P					
					Mix or Trial Number				
			Bulk		1	2	3	4	5
1. Coarse Aggregate	G_1		2.716	P_1			50.0		
2. Fine Aggregate	G_2		2.689	P_2			50.0		
3. Mineral Filler	G_3		--	P_3			--		
4. Total Aggregate	G_s	--	--	P_s			100.0		
5. Asphalt Cement	G_b	1.030	--	P_b			5.6		

	Equation*		1	2	3	4	5
6. Bulk Sp. Gr. (G_{sb}), total aggregate	(1)				2.703		
7. Max. Sp. Gr. (G_{mm}), paving mix ASTM D2041	—				2.535		
8. Bulk Sp. Gr. (G_{mb}), compacted mix ASTM D2726	—				2.442		
9. Effective Sp. Gr. (G_{se}), total aggregate	(2)				2.761		
10. Absorbed Asphalt (P_{ba}), % by wgt. total agg.	(4)				0.8		

CALCULATIONS

	Equation		1	2	3	4	5
11. Effective Asphalt Content (P_{be}) = Line 5 P_b – Line 10	(5)				4.5		
12. VMA = $100 - \dfrac{\text{Line 8}}{\text{Line 6}} \times \dfrac{100}{100 + \text{Line 5 } P_b} \times 100$	(7)				14.4		
13. Air Voids (V_a) = $100 \dfrac{\text{Line 7} - \text{Line 8}}{\text{Line 7}}$	(8)				3.7		
14. VFA = $100 \dfrac{\text{Line 12} - \text{Line 13}}{\text{Line 12}}$	(9)				74.3		

*Equations from Chapter 4, MS-2

Chapter 5

Marshall Method of Mix Design

A. General

5.01 DEVELOPMENT AND APPLICATION — The concepts of the Marshall method of designing paving mixtures were formulated by Bruce Marshall, a former Bituminous Engineer with the Mississippi State Highway Department. The U.S. Army Corps of Engineers, through extensive research and correlation studies, improved and added certain features to Marshall's test procedure, and ultimately developed mix design criteria. The Marshall test procedures have been standardized by the American Society for Testing and Materials. Procedures are given by ASTM D 1559, *Resistance to Plastic Flow of Bituminous Mixtures Using Marshall Apparatus.** Testing procedures presented here are basically the same as those of the ASTM method.

The original Marshall method is applicable only to hot-mix asphalt paving mixtures containing aggregates with maximum sizes of 25 mm (1 in.) or less. A modified Marshall method has been proposed for aggregates with maximum sizes up to 38 mm (1.5 in.). The differences between this proposed method and the original are discussed in Article 5.16. The Marshall method is intended for laboratory design and field control (Chapter 8) of asphalt hot-mix dense-graded paving mixtures. Because the Marshall stability test is empirical in nature, the meaning of the results in terms of estimating relative field behavior is lost when any modification is made to the standard procedures. An example of such modification is preparing specimens from reheated or remolded materials.

5.02 OUTLINE OF METHOD — The procedure for the Marshall method starts with the preparation of test specimens. Steps preliminary to specimen preparation are:
- (a) all materials proposed for use meet the physical requirements of the project specifications.
- (b) aggregate blend combinations meet the gradation requirements of the project specifications.
- (c) for performing density and voids analyses, the bulk specific gravity of all aggregates used in the blend and the specific gravity of the asphalt cement are determined.

---

* AASHTO T245 "Resistance to Plastic Flow of Bituminous Mixtures Using Marshall Apparatus" agrees with ASTM D 1559 except for provisions for mechanically-operated hammer. AASHTO T245 Par. 2.3 Note 2 - Instead of a hand-operated hammer and associated equipment, a mechanically-operated hammer may be used provided it has been calibrated to give results comparable to the hand-operated hammer.

These requirements are matters of routine testing, specifications, and laboratory technique that must be considered for any mix design method. Refer to Chapter 3, Evaluation of Aggregate Gradation, for the preparation and analysis of aggregates.

The Marshall method uses standard test specimens of 64 mm (2-1/2 in.) height x 102 mm (4 in.) diameter. These are prepared using a specified procedure for heating, mixing, and compacting the asphalt-aggregate mixture. The two principal features of the Marshall method of mix design are a density-voids analysis and a stability-flow test of the compacted test specimens.

The stability of the test specimen is the maximum load resistance in Newtons (lb.) that the standard test specimen will develop at 60°C (140°F) when tested as outlined. The flow value is the total movement or strain, in units of 0.25 mm (1/100 in.) occurring in the specimen between no load and the point of maximum load during the stability test.

B. Preparation of Test Specimens

5.03 GENERAL — In determining the design asphalt content for a particular blend or gradation of aggregates by the Marshall method, a series of test specimens is prepared for a range of different asphalt contents so that the test data curves show well-defined relationships. Tests should be planned on the basis of 1/2 percent increments of asphalt content, with at least two asphalt contents above the expected design value and at least two below this value.

The "expected design" asphalt content can be based on any or all of these sources: experience, computational formula, or performing the centrifuge kerosene equivalency and oil soak tests in the Hveem procedure (Chapter 6). Another quick method to arrive at a starting point is to use the dust-to-asphalt ratio guideline (0.6 to 1.2), discussed in Article 7.02. The expected design asphalt content, in percent by total weight of mix, could then be estimated to be approximately equivalent to the percentage of aggregate in the final gradation passing the 75 μ m (No. 200) sieve.

One example of a computational formula is this equation:

$$P = 0.035a + 0.045b + Kc + F$$

where:
- P = approximate asphalt content of mix, percent by weight of mix
- a = percent* of mineral aggregate retained on 2.36mm (No. 8) sieve
- b = percent* of mineral aggregate passing the 2.36mm (No. 8) sieve and retained on the 75 μ m (No. 200) sieve
- c = percent of mineral aggregate passing 75 μ m (No. 200) sieve
- K = 0.15 for 11-15 percent passing 75 μ m (No. 200) sieve
  - 0.18 for 6-10 percent passing 75 μ m (No. 200) sieve
  - 0.20 for 5 percent or less passing 75 μ m (No. 200) sieve
- F = 0 to 2.0 percent. Based on absorption of light or heavy aggregate. In the absence of other data, a value of 0.7 is suggested.

---

*Expressed as a whole number.

To provide adequate data, at least three test specimens are prepared for each asphalt content selected. Therefore, a Marshall mix design using six different asphalt contents will normally require at least eighteen test specimens. Each test specimen will usually require approximately 1.2 kg (2.7 lb) of aggregate. Assuming some minor waste, the minimum aggregate requirements for one series of test specimens of a given blend and gradation will be approximately 23 kg (50 lb). About four liters (one gallon) of asphalt cement will be adequate.

5.04 EQUIPMENT — The equipment required for the preparation of test specimens is:

(a) Flat-bottom metal *pans* for heating aggregates.
(b) *Round metal pans*, approximately 4-liter (4-qt.) capacity, for mixing asphalt and aggregate.
(c) *Oven and Hot Plate*, preferably thermostatically-controlled, for heating aggregates, asphalt, and equipment.
(d) *Scoop* for batching aggregates.
(e) *Containers*: gill-type tins, beakers, pouring pots, or sauce pans, for heating asphalt.
(f) *Thermometers*: armored, glass, or dial-type with metal stem, 10°C (50°F) to 235°C (450°F), for determining temperature of aggregates, asphalt and asphalt mixtures.
(g) *Balances*: 5-kg capacity, sensitive to 1 g, for weighing aggregates and asphalt and 2-kg capacity, sensitive to 0.1 g, for weighing compacted specimens.
(h) *Large Mixing Spoon* or small trowel.
(i) Large *spatula*.
(j) *Mechanical Mixer* (optional): commercial bread dough mixer 4-liter (4 qt.) capacity or larger, equipped with two metal mixing bowls and two wire stirrers.
(k) *Compaction Pedestal* (Figure 5.1), consisting of a 200 x 200 x 460 mm (8 x 8 x 18 in.) wooden post capped with a 305 x 305 x 25 mm (12 x 12 x 1 in.) steel plate. The wooden post should be oak, pine or other wood having a dry weight

Figure 5.1 – **Pedestal, hammer (mechanical) and mold used in preparing Marshall test specimens**

of 670 to 770 kg/m^3 (42 to 48 pcf). The wooden post should be secured by four angle brackets to a solid concrete slab. The steel cap should be firmly fastened to the post. The pedestal should be installed so that the post is plumb, the cap level, and the entire assembly free from movement during compaction.

*(l) *Compaction Mold*, consisting of a base plate, forming mold, and collar extension. The forming mold has an inside diameter of 101.6 mm (4 in.) and a height of approximately 75 mm (3 in.); the base plate and collar extension are designed to be interchangeable with either end of the forming mold.

*(m) *Compaction Hammer*, consisting of a flat circular tamping face, 98.4 mm (3-7/8 in.) in diameter and equipped with a 4.5 kg (10 lb.) weight constructed to obtain a specified 457 mm (18 in.) height of drop.

*(n) *Mold Holder*, consisting of spring tension device designed to hold compaction mold centered in place on compaction pedestal.

(o) *Paper disks*, 100mm (4 in.), for compaction.

(p) Steel specimen *extractor*, in the form of a disk with a diameter not less than 100 mm (3.95 in.) and 13 mm (0.5 in.) thick for extruding compacted specimens from mold.

(q) Welders *gloves* for handling hot equipment. Rubber gloves for removing specimens from water bath.

(r) *Marking Crayons*, for identifying test specimens.

(Note: See additional equipment requirements in Article 5.07.)

5.05 PREPARATION OF TEST SPECIMENS — These steps are recommended for preparing Marshall test specimens:

(a) *Number of Specimens* - Prepare at least three specimens for each combination of aggregates and asphalt content.

(b) *Preparation of Aggregates* - Dry aggregates to constant weight at 105°C to 110°C (220°F to 230°F) and separate the aggregates by dry sieving into the desired size fractions. These size fractions are recommended:

  25.0 to 19.0 mm (1 to 3/4 in.)
  19.0 to 9.5 mm (3/4 to 3/8 in.)
  9.5 to 4.75 mm (3/8 in. to No. 4)
  4.75 to 2.36 mm (No. 4 to No. 8)
  passing 2.36 mm (No. 8)

Refer to Article 3.13 for batching procedure details.

(c) *Determination of Mixing and Compaction Temperature* - The temperature to which the asphalt must be heated to produce viscosities of 170 ± 20 centistokes kinematic and 280 ± 30 centistokes kinematic shall be established as the mixing temperature and compaction temperatures, respectively. These temperatures can be estimated from a plot of the viscosity (log-log centistokes scale) versus temperature (log degrees Rankine scale, °R = °F + 459.7) relationship for the asphalt cement to be used. An example plot is shown in Figure 5.2.

---

*Marshall test apparatus should conform to requirements of ASTM D 1559.

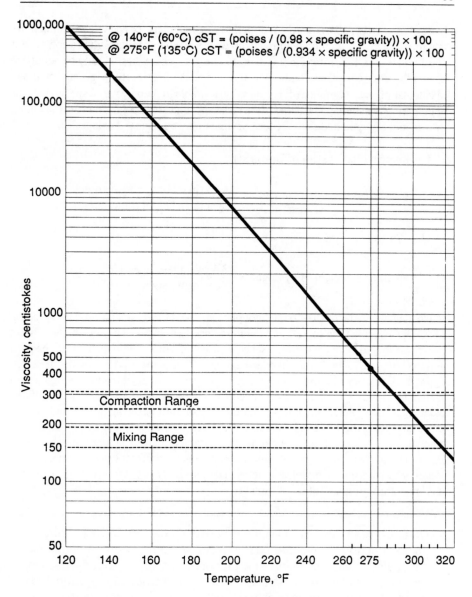

Figure 5.2 – **Determination of Mixing and Compaction Temperatures**

(d) *Preparation of Mold and Hammer* - Thoroughly clean the specimen mold assembly and the face of the compaction hammer and heat them in a water bath or on the hot plate to a temperature between 95°C and 150°C (200°F and 300°F). Place a piece of filter or waxed paper, cut to size, in the bottom of the mold before the mixture is placed in the mold.

(e) *Preparation of Mixtures* - Weigh into separate pans for each test specimen the amount of each size fraction required to produce a batch that will result in a compacted specimen 63.5 ± 1.27 mm (2.5 ± 0.05 in.) in height. This will normally be about 1.2 kg (2.7 lb.). (See Chapter 3, Evaluation of Aggregate Gradation, for suggested methods of calculating batch weights.) It is generally desirable to prepare a trial specimen prior to preparing the aggregate batches. If the trial specimen height falls outside the limits, the amount of aggregate used for the specimen may be adjusted using:

For International System of Units (SI),

$$\text{Adjusted mass of aggregate} = \frac{63.5 \text{ (mass of aggregate used)}}{\text{Specimen height (mm) obtained}}$$

U.S. Customary Units,

$$\text{Adjusted weight of aggregate} = \frac{2.5 \text{ (weight of aggregate used)}}{\text{Specimen height (in.) obtained}}$$

Place the pans in the oven or on the hot plate and heat to a temperature not exceeding 28°C (50°F) above the mixing temperature specified in (c). (If a hot plate is used, provision should be made for dead space, baffle plate, or a sand bath beneath the pans and the hot plate to prevent local over-heating.) Charge the mixing bowl with heated aggregates and dry mix thoroughly. Form a crater in the dry blended aggregate and weigh the required amount of asphalt cement into the mixture in accordance with the calculated batch weights. At this point the temperature of the aggregate and asphalt must be within the limits of the mixing temperature established in paragraph (c). Asphalt cement should not be held at mixing temperatures for more than one hour before using. Mix the aggregate and asphalt cement, preferably with a mechanical mixer or by hand with a trowel, as quickly and thoroughly as possible to yield a mixture having a uniform distribution of asphalt.

Note: Currently, there is no standardized or recommended procedure for aging or curing the mixture prior to Marshall compaction. A number of suggested methods have been proposed; however, a consensus of opinion has not yet been reached. The Hveem procedure recommends a 2 to 3 hour cure period to allow for both aging and absorption to occur. If severe climates or absorptive aggregates are involved, some consideration should be given to this behavior.

(f) *Packing the Mold* - Place a paper disk in the mold. Place the entire batch in the mold, spade the mixture vigorously with a heated spatula or trowel 15 times around the perimeter and ten times over the interior. Smooth the surface to a slightly rounded shape. The temperature of the mixture immediately prior to compaction shall be within the limits of the compaction temperature established in paragraph (c); otherwise, it shall be discarded. In no case shall the mixture be reheated.

(g) *Compaction of Specimens* - Place a paper on top of the mix and place the mold assembly on the compaction pedestal in the mold holder. As specified according to the design traffic category (see Table 5.2), apply either 35, 50, or 75 blows with the compaction hammer using a free fall of 457 mm (18 in.). Hold the axis of the compaction hammer as nearly perpendicular to the base of the mold assembly as possible during compaction. Remove the base plate and collar, and reverse and reassemble the mold. Apply the same number of compaction blows to the face of the reversed specimen. After compaction, remove the base plate and the paper disks and allow the specimen to cool in air until no deformation will result when removing it from the mold. When more rapid cooling is desired, electric fans may be used, but not water unless the specimen is in a plastic bag. Remove the specimen from the mold by means of an extrusion jack or other compression device, then place on a smooth, level surface until ready for testing. Normally, specimens are allowed to cool overnight.

Note: ASTM D1559 specifies that hand lifting of a flat faced compaction hammer be used for specimen compaction. If variations (e.g. mechanical lift, slanted face, rotating base) of the flat face, hand-lifted hammer are used, correlations with the standard Marshall compaction procedure must be made.

C. Test Procedure

5.06 GENERAL — In the Marshall method, each compacted test specimen is subjected to these tests and analysis in the order listed:
  (a) Bulk Specific Gravity Determination
  (b) Stability and Flow Test
  (c) Density and Voids Analysis

5.07 EQUIPMENT — The equipment required for the testing of the 102 mm (4 in.) diameter x 64 mm (2 1/2 in.) height specimens is:
 *(a) Marshall Testing Machine, a compression testing device. It is designed to apply loads to test specimens through cylindrical segment testing heads (inside radius of curvature of 51 mm (2 in.)) at a constant rate of vertical strain of 51 mm (2 in.) per minute. Two perpendicular guide posts are included to allow the two segments to maintain horizontal positioning and free vertical movement during the test. It is equipped with a calibrated proving ring for determining the applied testing load, a Marshall stability testing head for use in testing the specimen, and a Marshall flow meter for determining the amount of strain at the maximum load in the test. A universal testing machine equipped with suitable load and deformation indicating devices may be used instead of the Marshall testing frame.

*Marshall test apparatus should conform to requirements of ASTM D 1559.

(b) Water Bath, at least 150 mm (6 in.) deep and thermostatically-controlled to 60°C ± 1°C (140°F ± 1.8°F). The tank should have a perforated false bottom or be equipped with a shelf for suspending specimens at least 50 mm (2 in.) above the bottom of the bath.

5.08 BULK SPECIFIC GRAVITY DETERMINATION — The bulk specific gravity test may be performed as soon as the freshly-compacted specimens have cooled to room temperature. This test is performed according to ASTM D 1188, *Bulk Specific Gravity of Compacted Bituminous Mixtures Using Paraffin-coated Specimens* or ASTM D 2726, *Bulk Specific Gravity of Compacted Bituminous Mixtures Using Saturated Surface-Dry Specimens.*

5.09 STABILITY AND FLOW TESTS — After the bulk specific gravity of the test specimens have been determined, the stability and flow tests are performed:

(a) Immerse specimen in water bath at 60°C ± 1°C (140°F ± 1.8°F) for 30 to 40 minutes before test.

(b) If not using an automatic recording device(as shown in Figure 5.3), "Zero" the flow meter by inserting a 101.6 mm (4.00 in.) diameter metal cylinder in the testing head, placing the flow meter over the guide rod and adjusting the flow meter to read "zero."

(Note: This adjustment should be made on the guide post marked with an "O" and with the side of the upper segment of the testing head marked with an "O" being placed on the same side as the guide post so marked. The same assembly of testing head and flow meter must then be used in testing the specimens. Specimens should be 101.6 ± 0.25 mm [4.00 in. ± 0.01 in.]; otherwise, an initial and final reading of flow meter is required for the determination of the flow value.)

(c) Thoroughly clean the inside surfaces of testing head. Temperature of head shall be maintained between 21.1° to 37.8°C (70° to 100°F) using a water bath when required. Lubricate guide rods with a thin film of oil so that upper test head will slide freely without binding. If a proving ring is used to measure applied load, check to see that dial indicator is firmly fixed and "zeroed" for the "no-load" position.

(d) With testing apparatus ready, remove test specimen from water bath and carefully dry surface. Place specimen in lower testing head and center; then fit upper testing head into position and center complete assembly in loading device. Place flow meter over marked guide rod as noted in (b) above.

(e) Apply testing load to specimen at constant rate of deformation, 51 mm (2 in.) per minute, until failure occurs. The point of failure is defined by the maximum load reading obtained. The total number of Newtons (lb.) required to produce failure of the specimen shall be recorded as its Marshall stability value.

(f) While the stability test is in progress, if not using an automatic recording device, hold the flow meter firmly in position over guide rod and remove as the load begins to decrease, take reading and record. This reading is the flow value for the specimen, expressed in units of 0.25 mm (1/100 in.). For example, if the specimen deformed 3.8 mm (0.15 in.) the flow value is 15.

(g) The entire procedure for both the stability and flow measurements, starting with the removal of the specimen from the water bath, shall be completed within a period of thirty seconds.

Figure 5.3 – **Marshall stability and flow test, using an automatic recording device**

5.10 DENSITY AND VOIDS ANALYSIS — After the completion of the stability and flow test, a density and voids analysis is made for each series of test specimens. (The calculations for the voids analysis are fully described in Chapter 4.)

(a) Average the bulk specific gravity values for all test specimens of a given asphalt content; values obviously in error shall not be included in the average. These values of bulk specific gravity shall be used in further computations of voids data.

(b) Determine the average unit weight for each asphalt content by multiplying the average bulk specific gravity value by the density of water [1,000 kg/m^3 (62.4 pcf)] (See Figure 5.4).

(c) Determine the theoretical maximum specific gravity (ASTM D2041) for at least two asphalt contents, preferably on mixes at or near the design asphalt content. An average value for the effective specific gravity of the total aggregate is then calculated from these values. This value may then be used for calculation of the maximum specific gravity of mixtures with different asphalt contents, as discussed in Chapter 4.

(d) Using the effective and bulk specific gravity of the total aggregate, the average bulk specific gravities of the compacted mix, the specific gravity of the asphalt, and the maximum specific gravity of the mix determined above in (c), calculate the percent absorbed asphalt by weight of dry aggregate, percent air voids (V_a), percent voids filled with asphalt (VFA) and percent voids in mineral aggregate (VMA). These values and calculations are more fully described in Chapter 4. Worksheets for all these calculations are also included in Chapter 4.

D. Interpretation of Test Data

5.11 PREPARATION OF TEST DATA — Prepare the stability and flow values and void data:

(a) Measured stability values for specimens that depart from the standard 63.5 mm (2 1/2 in.) thickness shall be converted to an equivalent 63.5 mm (2 1/2 in.) value by means of a conversion factor. Applicable correlation ratios to convert the measured stability values are set forth in Table 5.1. Note that the conversion may be made on the basis of either measured thickness or measured volume.

(b) Average the flow values and the final converted stability values for all specimens of a given asphalt content. Values that are obviously in error shall not be included in the average.

(c) Prepare a separate graphical plot for these values and connect plotted points with a smooth curve that obtains the "best fit" for all values, as illustrated in Figure 5.5:

Stability vs. Asphalt Content
Flow vs. Asphalt Content
Unit Weight of Total Mix vs. Asphalt Content
Percent Air Voids (V_a) vs. Asphalt Content
Percent Voids Filled with Asphalt (VFA) vs. Asphalt Content
Percent Voids in Mineral Aggregate (VMA) vs. Asphalt Content

These graphs are used to determine the design asphalt content of the mix.

Compaction: 75 Blows
Specific Gravity of AC: 1.030
Bulk S.G. Aggregate: 2.674

Grade AC: AC-20
Absorbed AC of Aggregate: 0.6%
Effective S.G. Aggregate: 2.717

Project:
Location:
Date:

Trial Mix:

% AC by wt. of mix, Spec. No.	Spec. Height in. (mm)	Mass, grams — In Air	In Water	Sat. Surface Dry In Air	Bulk Volume, cc	Bulk S.G. Specimen	Max. S.G. (Loose Mix)	Unit Weight, pcf (Mg/m³)	% Air Voids	% VMA	% VFA	Stability, lbs (N) Measured	Adjusted	Flow 0.01 in. (0.25mm)
3.5 - A		1240.6	726.4	1246.3	519.9	2.386		148.9				2440	2440	8
3.5 - B		1238.7	723.3	1242.6	519.3	2.385		148.8				2420	2420	7
3.5 - C		1240.1	724.1	1245.9	521.8	2.377		148.3				2510	2510	7
Average						2.383	2.570	148.7	7.3	14.0	48.0		2457	7
4.0 - A		1244.3	727.2	1246.6	519.4	2.396		149.5				2180	2180	9
4.0 - B		1244.6	727.0	1247.6	520.6	2.391		149.2				2260	2260	9
4.0 - C		1242.6	727.9	1244.0	516.1	2.408		150.2				2310	2310	8
Average						2.398	2.550	149.6	6.0	13.9	57.1		2250	9
4.5 - A		1249.3	735.8	1250.2	514.4	2.429		151.5				2420	2420	9
4.5 - B		1250.8	728.1	1251.6	523.5	2.389		149.1				2410	2314	9
4.5 - C		1251.6	735.3	1253.1	517.8	2.417		150.8				2340	2340	9
Average						2.412	2.531	150.5	4.7	13.9	66.1		2358	9
5.0 - A		1256.7	739.8	1257.6	517.8	2.427		151.4				2290	2290	9
5.0 - B		1258.7	742.7	1259.3	516.6	2.437		152.0				2190	2190	8
5.0 - C		1258.4	737.5	1259.1	521.6	2.413		150.5				2240	2240	9
Average						2.425	2.511	151.3	3.4	13.8	75.2		2240	9
5.5 - A		1263.8	742.6	1264.3	521.7	2.422		151.2				2210	2210	9
5.5 - B		1258.8	741.4	1259.4	518.0	2.430		151.6				2300	2300	10
5.5 - C		1259.0	742.5	1259.5	517.0	2.435		152.0				2210	2210	9
Average						2.429	2.493	151.6	2.5	14.1	82.1		2240	9

Figure 5.4 – Suggested test report form showing test data for a typical design by the Marshall method

Table 5.1 – **Stability correlation ratios**

Volume of Specimen, cm³	Approximate Thickness of Specimen, mm	in.	Correlation Ratio
200 to 213	25.4	1	5.56
214 to 225	27.0	1 1/16	5.00
226 to 237	28.6	1 1/8	4.55
238 to 250	30.2	1 3/16	4.17
251 to 264	31.8	1 1/4	3.85
265 to 276	33.3	1 5/16	3.57
277 to 289	34.9	1 3/8	3.33
290 to 301	36.5	1 7/16	3.03
302 to 316	38.1	1 1/2	2.78
317 to 328	39.7	1 9/16	2.50
329 to 340	41.3	1 5/8	2.27
341 to 353	42.9	1 11/16	2.08
354 to 367	44.4	1 3/4	1.92
368 to 379	46.0	1 13/16	1.79
380 to 392	47.6	1 7/8	1.67
393 to 405	49.2	1 15/16	1.56
406 to 420	50.8	2	1.47
421 to 431	52.4	2 1/16	1.39
432 to 443	54.0	2 1/8	1.32
444 to 456	55.6	2 3/16	1.25
457 to 470	57.2	2 1/4	1.19
471 to 482	58.7	2 5/16	1.14
483 to 495	60.3	2 3/8	1.09
496 to 508	61.9	2 7/16	1.04
509 to 522	63.5	2 1/2	1.00
523 to 535	65.1	2 9/16	0.96
536 to 546	66.7	2 5/8	0.93
547 to 559	68.3	2 11/16	0.89
560 to 573	69.8	2 3/4	0.86
574 to 585	71.4	2 13/16	0.83
586 to 598	73.0	2 7/8	0.81
599 to 610	74.6	2 15/16	0.78
611 to 625	76.2	3	0.76

NOTES:
1. The measured stability of a specimen multiplied by the ratio for the thickness of the specimen equals the corrected stability for a 63.5 mm (2 1/2-in.) specimen.
2. Volume-thickness relationship is based on a specimen diameter of 101.6 mm (4 in.).

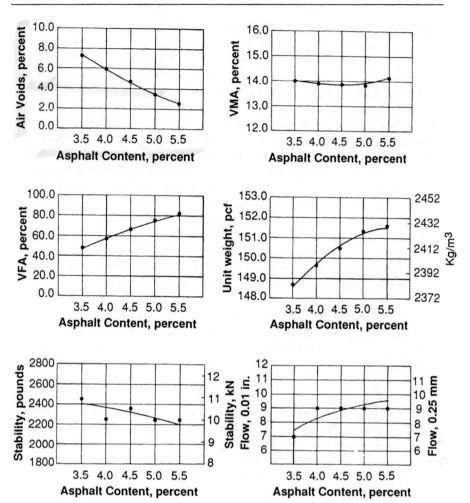

Figure 5.5 – **Test property curves for hot-mix design data by the Marshall method**

5.12 TRENDS AND RELATIONS OF TEST DATA — By examining the test property curves plotted for Article 5.11, information can be learned about the sensitivity of the mixture to asphalt content. The test property curves have been found to follow a reasonably consistent pattern for dense-graded asphalt paving mixes, but variations will and do occur. Trends generally noted are:

(a) The stability value increases with increasing asphalt content up to a maximum after which the stability decreases.

(b) The flow value consistently increases with increasing asphalt content.

(c) The curve for unit weight of total mix follows the trend similar to similar to the stability curve, except that the maximum unit weight normally (but not always) occurs at a slightly higher asphalt content than the maximum stability.

(d) The percent of air voids, V_a, steadily decreases with increasing asphalt content, ultimately approaching a minimum void content.

(e) The percent voids in the mineral aggregate, VMA, generally decreases to a minimum value then increases with increasing asphalt content.

(f) The percent voids filled with asphalt, VFA, steadily increases with increasing asphalt content, because the VMA is being filled with asphalt.

5.13 CRITERIA FOR SATISFACTORY PAVING MIX — Deciding whether the asphalt paving mix will be satisfactory at the selected design asphalt content is guided by applying certain limiting criteria to the mixture test data. The Marshall method mix design criteria in Table 5.2 are recommended by the Asphalt Institute.

5.14 DETERMINATION OF PRELIMINARY DESIGN ASPHALT CONTENT — The design asphalt content of the asphalt paving mix is selected by considering all of the data discussed previously. As an initial starting point, the Asphalt Institute recommends choosing the asphalt content at the median of the percent air voids limits, which is four percent. All of the calculated and measured mix properties at this asphalt content should then be evaluated by comparing them to the mix design criteria in Table 5.2. If all of the criteria are met, then this is the preliminary design asphalt content. If all of the design criteria are not met, then some adjustment or compromise is necessary or the mix may need to be redesigned. A number of considerations are discussed in the next article that should be weighed even if all the design criteria are met.

EXAMPLE

Assume the data shown in Figures 5.4 and 5.5 represent Marshall mix design laboratory tests on dense-graded HMA to be used in a heavy traffic area. The mixture contains a 3/4" nominal maximum particle size aggregate. At four percent air voids, the mixture properties are:

Asphalt Content, %	4.7
Stability, lb.	2,300
Flow, 0.01 in.	9
VMA, %	14
VFA, %	70

Table 5.2 – **Marshall mix design criteria**

Marshall Method Mix Criteria [1]	Light Traffic Surface & Base		Medium Traffic Surface & Base		Heavy Traffic Surface & Base	
	Min	Max	Min	Max	Min	Max
Compaction, number of blows each end of specimen	35		50		75	
Stability, N (lb.)	3336 (750)	—	5338 (1200)	—	8006 (1800)	—
Flow, 0.25 mm (0.01 in.)	8	18	8	16	8	14
Percent Air Voids	3	5	3	5	3	5
Percent Voids in Mineral Aggregate (VMA)			See Table 5.3			
Percent Voids Filled With Asphalt (VFA)	70	80	65	78	65	75

NOTES
1. All criteria, not just stability value alone, must be considered in designing an asphalt paving mix. Hot mix asphalt bases that do not meet these criteria when tested at 60°C (140°F) are satisfactory if they meet the criteria when tested at 38°C (100°F) and are placed 100 mm (4 inches) or more below the surface. This recommendation applies only to regions having a range of climatic conditions similar to those prevailing throughout most of the United States. A different lower test temperature may be considered in regions having more extreme climatic conditions.
2. Traffic classifications
   Light   Traffic conditions resulting in a Design EAL <10^4
   Medium Traffic conditions resulting in a Design EAL between 10^4 and 10^6
   Heavy   Traffic conditions resulting in a Design EAL >10^6
3. Laboratory compaction efforts should closely approach the maximum density obtained in the pavement under traffic.
4. The flow value refers to the point where the load begins to decrease.
5. The portion of asphalt cement lost by absorption into the aggregate particles must be allowed for when calculating percent air voids.
6. Percent voids in the mineral aggregate is to be calculated on the basis of the ASTM bulk specific gravity for the aggregate.

Comparing these values to the criteria in Table 5.2, it is evident that this mixture is acceptable for use in heavy traffic areas.

5.15 SELECTION OF FINAL MIX DESIGN — The final selected mix design is usually the most economical one that will satisfactorily meet all of the established criteria. However, the mix should not be designed to optimize one particular property. Mixes with abnormally high values of stability are often less desirable because pavements with such mixes tend to be less durable and may crack prematurely under heavy volumes of traffic. This situation is especially critical where the base and subgrade materials beneath the pavement are weak and permit moderate to relatively high deflections under the actual traffic.

The design asphalt content should be a compromise selected to balance all of the mix properties. Normally, the mix design criteria will produce a narrow range of acceptable asphalt contents that pass all of the guidelines as shown by the example in Figure 5.6. The asphalt content selection can be adjusted <u>within this narrow range</u> to achieve a mix

Table 5.3 – **Minimum percent voids in mineral aggregate (VMA)**

Nominal Maximum Particle Size[1,2]		Minimum VMA, percent		
		Design Air Voids, Percent[3]		
mm	in.	3.0	4.0	5.0
1.18	No. 16	21.5	22.5	23.5
2.36	No. 8	19.0	20.0	21.0
4.75	No. 4	16.0	17.0	18.0
9.5	3/8	14.0	15.0	16.0
12.5	1/2	13.0	14.0	15.0
19.0	3/4	12.0	13.0	14.0
25.0	1.0	11.0	12.0	13.0
37.5	1.5	10.0	11.0	12.0
50	2.0	9.5	10.5	11.5
63	2.5	9.0	10.0	11.0

1 - Standard Specification for Wire Cloth Sieves for Testing Purposes, ASTM E11 (AASHTO M92)
2 - The nominal maximum particle size is one size larger than the first sieve to retain more than 10 percent.
3 - Interpolate minimum voids in the mineral aggregate (VMA) for design air void values between those listed.

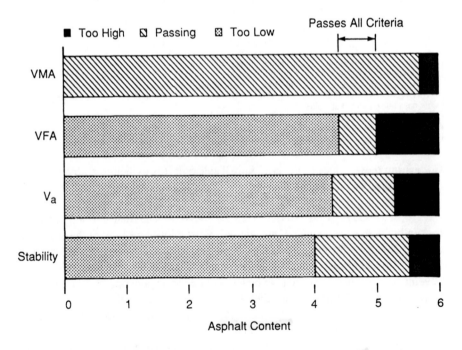

Figure 5.6 – **An example of the narrow range of acceptable asphalt contents.**

property that will satisfy a requirement of a specific project. Different properties are more critical for different circumstances, depending on traffic, structure, climate, construction equipment, and other factors. Therefore, the balancing process is not the same for every pavement and every mix design. These are some considerations for adjustment that should be evaluated prior to establishing the final design asphalt content:

Evaluation of VMA Curve

In many cases, the most difficult mix design property to achieve is a minimum amount of voids in the mineral aggregate. The goal is to furnish enough space for the asphalt cement so it can provide adequate adhesion to bind the aggregate particles, but without bleeding when temperatures rise and the asphalt expands. Normally, the curve exhibits a flattened U-shape, decreasing to a minimum value and then increasing with increasing asphalt content, shown in Figure 5.7(a).

This dependency of VMA on asphalt content appears to be a contradiction to the definition. One might expect the VMA to remain constant with varying asphalt content, thinking that the air voids would simply be displaced by asphalt cement. In reality, the total volume changes across the range of asphalt contents; the assumption of a constant unit volume is not accurate. With the increase in asphalt, the mix actually becomes more workable and compacts more easily, meaning more weight can be compressed into less volume. Therefore, up to a point, the bulk density of the mix increases and the VMA decreases.

At some point as the asphalt content increases (the bottom of the U-shaped curve) the VMA begins to increase because relatively more dense material (aggregate) is displaced and pushed apart by the less dense material (asphalt cement). It is recommended that <u>asphalt contents on the "wet" or right-hand increasing side of this VMA curve be avoided</u>, even if the minimum air void and VMA criteria is met. Design asphalt contents in this range have a tendency to bleed and/or exhibit plastic flow when placed in the field. Any amount of additional compaction from traffic leads to inadequate room for asphalt expansion, loss of aggregate-to-aggregate contact, and eventually, rutting and shoving in high traffic areas. Ideally, the design asphalt content should be selected slightly to the left of the low point of the VMA curve, provided none of the other mixture criteria are violated.

In some mixes, the bottom of the U-shaped VMA curve is very flat, meaning that the compacted mixture is not as sensitive to asphalt content in this range as some other factors. In the normal range of asphalt contents, compactability is influenced more by aggregate properties. However, at some point the quantity of asphalt will become critical to the behavior of the mix and the effect of asphalt will dominate as the VMA increases drastically.

When the bottom of the U-shaped VMA curve falls below the minimum criteria level required for the nominal maximum aggregate size of the mix [Figure 5.7(b)], this is an indication that changes to the job-mix formula are necessary. Specifically, the aggregate grading should be modified to provide additional VMA; suggestions are provided in Chapter 3. <u>The design asphalt content should not be selected at the extremes of the acceptable range even though the minimum criteria are met.</u> On the

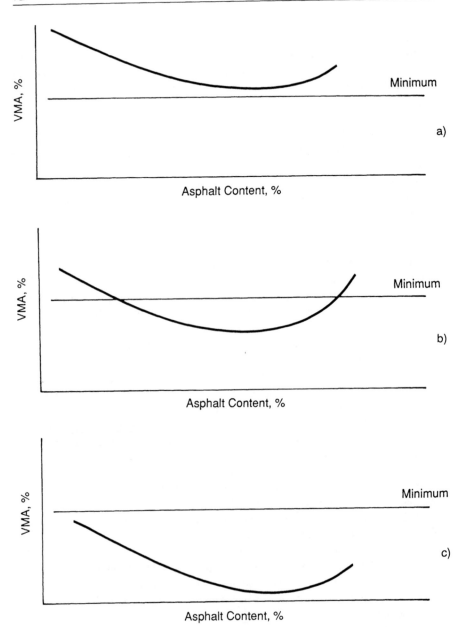

Figure 5.7 – **Relationship between VMA and specification limit**

left-hand side, the mix would be too dry, prone to segregation, and would probably be too high in air voids. On the right-hand side, the mix would be expected to rut.

If the minimum VMA criteria is completely violated over the entire asphalt content range [curve is completely below minimum, Figure 5.7(c)], a significant redesign and/or change in material sources is warranted.

Effect of Compaction Level

At the same asphalt content, both air voids (V_a) and voids in the mineral aggregate (VMA) decrease with higher compactive effort. The three levels of compaction of the Marshall mix procedure can be used to illustrate the consequences of this fact. As shown in Figure 5.8(a), not only do the magnitudes of the values change but the asphalt content value at the minimum VMA shifts. If a mix is designed slightly to the left of minimum VMA at a compaction level of 50 blows and the pavement actually endures heavier traffic than expected (closer to 75-blow design level); then, the same asphalt content now plots on the right-hand or "wet" side of the minimum VMA point for a mix designed using 75 blow compaction. Ultimately, a mix susceptible to rutting is the result.

This scenario can also work in the opposite direction. If a mix, designed at a compaction level of 75 blows as shown in Figure 5.8(b), is placed in a pavement with much lower volumes of traffic, then the final percentage of air voids (V_a) will be considerably higher than planned. This condition could lead to a more open, permeable mix allowing air and water to pass through easily. The consequence of this situation is a mix that hardens prematurely, becomes brittle and cracks at an early age or the aggregate ravels out of the mix because of the loss of asphalt adhesion. This condition may also lead to stripping as discussed in Chapter 7.

For this reason, it is important that the compactive effort used to simulate the design traffic expected in the pavement be selected accordingly in the laboratory. Also, the mixture must be constructed with appropriate compaction equipment in the field to produce adequate initial density regardless of climatic conditions.

It is also important to note that the VMA criteria do not change based on the level of compaction. The reasoning for having sufficient VMA (providing space for the asphalt and air voids) is consistent regardless of the traffic level for which the mixture is being designed.

Effect of Air Voids

It should be emphasized that the design range of air voids (3 to 5 percent) is the level desired after several years of traffic. This goal does not vary with traffic as seen in Table 5.2; the laboratory compactive effort is supposed to be selected for the expected traffic. This design air void range will normally be achieved if the mix is designed at the correct compactive effort and the percent air voids after construction is about 8 percent. Some consolidation with traffic is expected and desired.

The consequence of a change in any factor or any detour in the procedure that offsets the total process will be a loss of performance or service life. It has been shown that mixtures that ultimately consolidate to less than three percent air voids can be expected to rut and shove if placed in heavy traffic locations. Several factors may contribute to this occurrence, such as an arbitrary or accidental increase in asphalt content at the

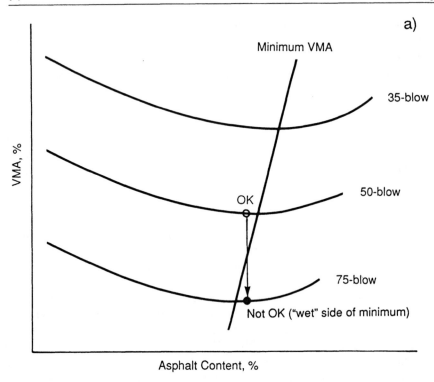

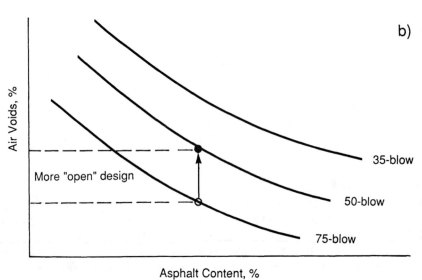

Figure 5.8 – **Effect of Marshall compactive effort on VMA and air voids.**

mixing facility or an increased amount of ultra-fine particles passing the 75 μ m (No. 200) sieve beyond that used in the laboratory, which will act as an asphalt extender.

Similarly, problems can occur if the final air void content is above five percent or if the pavement is constructed with over eight percent air voids initially. Brittleness, premature cracking, ravelling, and stripping are all possible under these conditions.

The overall objective is to limit adjustments of the design asphalt content to less than 0.5 percent air voids from the median of the design criteria (four percent), especially on the low side of the range, and to verify that the plant mix closely resembles the laboratory mix.

Effect of Voids Filled with Asphalt

Although VFA, VMA, and V_a are all interrelated and only two of the values are necessary to solve for the other, including the VFA criteria helps prevent the design of mixes with marginally-acceptable VMA. The main effect of the VFA criteria is to limit maximum levels of VMA, and, subsequently, maximum levels of asphalt content.

VFA also restricts the allowable air void content for mixes that are near the minimum VMA criteria. Mixes designed for lower traffic volumes will not pass the VFA criteria with a relatively high percent air voids (five percent) even though the air void criteria range is met. The purpose is to avoid less durable mixes in light traffic situations.

Mixes designed for heavy traffic will not pass the VFA criteria with relatively low percent air voids (less than 3.5 percent) even though that amount of air voids is within the acceptable range. Because low air void contents can be very critical in terms of permanent deformation (as discussed previously), the VFA criteria helps to avoid those mixes that would be susceptible to rutting in heavy traffic situations.

The VFA criteria provide an additional factor of safety in the design and construction process in terms of performance. Since changes can occur between the design stage and actual construction, an increased margin for error is desirable.

Influence of Structure and Climate

Mix design is a compromise of many factors. The asphalt content that provides the best overall performance, in addition to passing the previously-discussed conventional criteria, would be considered the design value. The CAMAS computer program, contained in the Asphalt Institute *Computer-Assisted Asphalt Mix Analysis System* package, provides an additional tool for evaluating the predicted performance of a specific mix placed in a particular situation. The various mathematical models contained in the program have not been fully verified and the program is currently considered only a research tool. However, models are included for examining fatigue life, subgrade deformation, and asphalt concrete deformation of the pavement and mix for the actual climatic and traffic conditions. If any of these levels of performance are not acceptable, then either the mix or the structure could be modified and a subsequent evaluation performed.

The decision-making process for selecting the design asphalt content in the mix varies with the circumstances involved in the specific case. Depending on the particular structure or agency policy, certain factors may be more important than others. Although it was found that it is not feasible, suitable, or practical to directly trade-off thickness for better mix compaction or a change in asphalt content, there are other advantages to integrating structural design and mix design.

The type of structure can alter the engineer's evaluation in many ways. For example, in an asphalt concrete overlay of a portland cement concrete pavement, there would be little concern for fatigue, since the tensile strains in the bottom of the AC overlay would be minimal. This is also true for the subgrade deformation related to the compressive strain on the top of the subgrade. The main consideration would be how to limit the AC rutting as well as any supplementary treatments for minimizing reflective cracking. In this particular case, it may be worthwhile to look at the effects of altering the compactive effort in the lab and field while changing the asphalt content. Depending on the environmental conditions during construction, it may be worthwhile to use heavier rollers or a longer period of rolling to achieve more or the same density with less asphalt in the mix. Mixes with asphalt contents on the high-side of the acceptable range are usually avoided in this situation.

In an asphalt pavement, all three performance indicators need to be evaluated in terms of future maintenance. Initially, it is important that the subgrade be adequately protected by the structure; the number of allowable repetitions based on subgrade deformation should exceed that expected or the pavement's performance may have little to do with proper mix design. In some cases, depending on location and traffic volume, the engineer may consider whether cracking or rutting is less of a future maintenance concern and the mix design can be selected accordingly. With all other factors being equal, mixes with asphalt contents on the high-side of the range are less prone to cracking because of the additional flexibility. Similarly, mixes on the low-side of the range are less susceptible to rutting.

Finally, climate can have a major impact on mix and pavement performance for a given pavement structure. Mix designs do not usually consider this factor except in selecting the category or grade of asphalt cement. Table 5.4 gives recommended asphalt grades for various temperature conditions.

In hot climates, harder, more viscous asphalts are normally used to obtain more stability from asphalt adhesion as well as from aggregate interlock. If the mix is designed and constructed to maximize aggregate-to-aggregate contact, then the properties of the asphalt cement are less important. Regardless, asphalt contents on the low-side of the acceptable range are recommended for these areas.

In colder climates, softer, less viscous asphalts are recommended to produce a mix which is less susceptible to low-temperature shrinkage cracking. Rutting is less of a concern; therefore, additional stability from asphalt adhesion is not necessary. Usually, asphalt contents on the high-side of the acceptable range are recommended to furnish a mix which is more elastic and resilient.

Specific Project Conditions

The season of the year when the pavement is being constructed can be another factor to be considered when selecting the final design asphalt content. Summer paving would usually call for lower asphalt contents, while fall or early spring construction would dictate higher asphalt contents to assist compaction in cooler temperatures. Any

Table 5.4 – **Selecting Asphalt Grade**

Temperature Condition	Asphalt Grades	
Cold, mean annual air temperature ≤7°C (45°F)	AC-5, AR-2000, 120/150 pen.	AC-10 AR-4000 85/100 pen.
Warm, mean annual air temperature between 7°C (45°F) and 24°C (75°F)	AC-10, AR-4000, 85/100 pen.	AC-20 AR-8000 60/70 pen.
Hot, mean annual air temperature ≥24°C (75°F)	AC-20, AR-8000, 60/70 pen.	AC-40 AR-16000 40/50 pen.

shift in asphalt content is only a minor amount within the narrow range that passes all the previous criteria.

The amount and handling of traffic can also influence the final decision. If the actual traffic is at the low or high end of the broad traffic categories for selecting the laboratory compactive effort and mix design criteria, then the asphalt content could be slightly modified accordingly. Higher traffic areas would demand the lower asphalt contents within the acceptable range. Mixes to be used in overlay situations with reduced lane detours, where the pavement will undergo severe loading concentrations such as highly-channelized wheel passes, very slow speeds, or steep upgrades, demand additional attention in all phases of production. The design asphalt content should be selected from the low end of the acceptable range and initial compaction requirements must be met. Traffic should be held off of the pavement as long as possible while the mix is cooling to normal temperatures. This cooling allows the asphalt to contribute more to the mix stability and less as a compaction lubricant.

5.16 MODIFIED MARSHALL METHOD FOR LARGE AGGREGATE — A modified Marshall method has been developed by Kandhal of the National Center for Asphalt Technology for mixes composed of aggregates with maximum size up to 38 mm (1.5 in.). This procedure is documented in draft form in the 1990 Proceedings of the Association of Asphalt Paving Technologists (AAPT). The procedure is basically the same as the original method except for these differences that are due to the larger specimen size that is used:

(a) The hammer weighs 10.2 kg (22.5 lb.) and has a 149.4 mm (5.88 in.) flat tamping face. Only a mechanically-operated device is used for the same 457 mm (18 in.) drop height.

(b) The specimen has a 152.4 mm (6 in.) diameter by 95.2 mm (3.75 in.) height.

(c) The batch weights are typically 4 kg.

(d) The equipment for compacting and testing (molds and breaking heads) are proportionately larger to accommodate the larger specimens.

(e) The mix is placed in the mold in two approximately equal increments, with spading performed after each increment to avoid honey-combing.

(f) The number of blows needed for the larger specimen is 1.5 times (75 or 112 blows) that required of the smaller specimen (50 or 75 blows) to obtain equivalent compaction.

(g) The design criteria should be modified as well. The minimum stability should be 2.25 times and the range of flow values should be 1.5 times the same criteria listed in Table 5.2 for the normal-sized specimens.

(h) Similar to the normal procedure, these values should be used to convert the measured stability values to an equivalent value for a specimen with a 95.2 mm (3.75 in.) thickness, if the actual thickness varies:

Approximate Height		Specimen Volume	Correlation
(mm)	(in.)	(cc)	Ratio
88.9	3 1/2	1608 to 1626	1.12
90.5	3 9/16	1637 to 1665	1.09
92.1	3 5/8	1666 to 1694	1.06
93.7	3 11/16	1695 to 1723	1.03
95.2	3 3/4	1724 to 1752	1.00
96.8	3 13/16	1753 to 1781	0.97
98.4	3 7/8	1782 to 1810	0.95
100.0	3 15/16	1811 to 1839	0.92
101.6	4	1840 to 1868	0.90

Chapter 6

Hveem Method of Mix Design

A. General

6.01 DEVELOPMENT AND APPLICATION — The concepts of the Hveem method of designing paving mixtures have been advanced and developed under the direction of Francis N. Hveem, a former Materials and Research Engineer for the California Department of Transportation. The Hveem method has been developed over a period of years as certain features have been improved and other features added. The test procedures and their application have been developed through extensive research and correlation studies on asphalt highway pavements.

The Hveem method as developed and used by the California Department of Transportation and others is applicable to paving mixtures using either asphalt cement or cutback asphalt and containing aggregates up to 25 mm (1 in.) maximum size. The method presented here is applicable to the design of hot asphalt, dense-graded paving mixtures.

Hveem method test procedures have been standardized by the American Society for Testing and Materials. Test procedures are found in ASTM D 1560, *Resistance to Deformation and Cohesion of Bituminous Mixtures by Means of Hveem Apparatus,* and ASTM D 1561, *Preparation of Bituminous Mixture Test Specimens by Means of California Kneading Compactor.* Testing procedures presented in this manual are basically the same as those of the ASTM test methods.

6.02 OUTLINE OF METHOD — The procedure for the Hveem method begins with the preparation of test specimens. Steps preliminary to specimen preparation are:
- (a) proposed materials meet the physical requirements of the project specifications.
- (b) aggregate blend combinations meet the gradation requirements of the project specifications.
- (c) An ample supply of aggregates is dried and sized into fractions.

These requirements are matters of routine testing, specifications, and laboratory technique, which must be considered but are not unique to any particular mix design method. The reader should refer to Chapter 3 for the schedule of preparation and analysis of aggregates. It should be noted, however, that the maximum size aggregates used in the test mixes should not exceed 25 mm (1 in.). In the event that the specifications for the paving mix being considered require aggregate sizes greater than 25 mm (1 in.), up to 25 percent of oversized rock may be screened out. However, this screening process can have a significant effect on the magnitude of the stabilometer values, depending on the size, amount, and shape of the larger aggregate pieces.

The Hveem method uses standard test specimens of 64 mm (2 1/2 in.) height by 102 mm (4 in.) diameter; these are prepared using a specified procedure for heating, mixing, and compacting the asphalt-aggregate mixtures. The principal features of the Hveem method of mix design are the surface capacity and Centrifuge Kerosene Equivalent (C. K. E.) test on the aggregates to estimate the asphalt requirements of the mix, followed by a stabilometer test, a cohesiometer test,* a swell test, and a density voids analysis on test specimens of the compacted paving mixtures. The stabilometer test utilizes a special triaxial-type testing cell for measuring the resistance of the compacted mix to lateral displacement under vertical loading, and the swell test measures the resistance of the mix to the action of water. The specimens are maintained at 60°C (140°F) for the stability test, and the swell test is performed at room temperature.

B. Approximate Asphalt Content by the Centrifuge Kerosene Equivalent Method

6.03 GENERAL — The first step in the Hveem method of mix design is to determine the "approximate" asphalt content by the Centrifuge Kerosene Equivalent method.** With a calculated surface area and the factors obtained by the C.K.E. method for a particular aggregate or blend of aggregates, the approximate asphalt content is determined by using a series of charts. These charts are presented in this chapter, accompanied by typical examples to demonstrate their application.

6.04 EQUIPMENT — The equipment and materials required for determining the approximate asphalt content (Figure 6.1) are:
  (a) Small *sample splitter* for obtaining representative samples of fine aggregate.
  (b) *Pans,* 114 mm (4 1/2 in.) diameter x 25 mm (1 in.) deep.
  (c) *Kerosene,* 4 liters (1 gal).
  (d) *Oil,* SAE No 10, lubricating. 4 liters (1 gal)
  (e) *Beakers,* 1500 ml.
  (f) *Metal Funnels,* 89 mm (3 1/2 in.) top diameter, 114 mm (4 1/2 in.) height, 13 mm (1/2 in.) orifice with piece of 2.00 mm (No. 10) sieve soldered to bottom of opening.
  (g) *Timer.*
  (h) *Centrifuge,* hand-operated, complete with cups, capable of producing 400 times gravity (a power-driven centrifuge is available from Soiltest, Inc., Lake Bluff, Illinois , Catalog No. AP-275 or equivalent.)
  (i) *Filter Papers,* 55 mm diam. (No. 611, Eaton-Dikeman Co., Mt. Holly Springs, Pennsylvania, or equivalent).
(Note: See Articles 6.10 and 6.15 for additional Hveem method equipment requirements.)

---

*The cohesiometer value is seldom used and, therefore, this test is not included in this manual. If the cohesiometer value is desired the test is described in ASTM D 1560.

**The development of the method for determining design asphalt content is outlined in "Establishing the Oil Content for Dense-Graded Bituminous Mixtures" by F. N. Hveem, California Highways and Public Works, July-August, 1942, and also presented in the Proceedings of The Association of Asphalt Paving Technologists, Volume 13, 1942.

Figure 6.1 – **Apparatus for Hveem C.K.E. tests**

6.05 SURFACE AREA — The gradation of the aggregate or blend of aggregates employed in the mix is used to calculate the surface area of the total aggregate. This calculation consists of multiplying the total percent passing each sieve size by a "surface-area factor" as set forth in Table 6.1. Sum these products and the total will represent the equivalent surface area of the sample in terms of m^2/kg (ft^2/lb.). It is important to note that all the surface-area factors must be used in the calculation. Also, if a different series of sieves is used, different surface-area factors are necessary.

Note: These surface-area factors have been used to calculate an average film thickness using the volume of asphalt binder in the mix. Although this determination of asphalt film thickness can provide a broad, relative indication of mix durability, the Asphalt Institute strongly recommends against comparing this calculated value with specific mix design criteria because of inherent inaccuracies. These surface-area factors do not take into account the specific aggregate shape, but are intended only as an index factor. In addition, in a compacted mixture some of the asphalt and fine particle mastic is actually shared by adjacent particles rather then each being in an isolated state as assumed.

Table 6.1 – **Surface area factors**

Total Percent Passing Sieve No.	Maximum Size	4.75 mm (No. 4)	2.36 mm (No. 8)	1.18 mm (No. 16)	600 μm (No. 30)	300 μm (No. 50)	150 μm (No. 100)	75 μm (No. 200)
Surface-Area Factor,* m2/kg (ft2/lb.)	.41 (2)	.41 (2)	.82 (4)	1.64 (8)	2.87 (14)	6.14 (30)	12.29 (60)	32.77 (160)

*Surface area factors shown are applicable only when all the above-listed sieves are used in the sieve analysis.

This example tabulation demonstrates the calculation of surface area by this method.

Sieve Size	Percent Passing	×	S.A. Factor m2/kg (ft2/lb)	=	Surface Area m2/kg (ft2/lb)
19.0 mm (3/4 in.)	100 }*		.41 (2)		.41 (2)
9.5 mm (3/8 in.)	90				
4.75 mm (No. 4)	75		.41 (2)		.31 (1.5)
2.36 mm (No. 8)	60		.82 (4)		.49 (2.4)
1.18 mm (No. 16)	45		1.64 (8)		.74 (3.6)
600 μm (No. 30)	35		2.87 (14)		1.00 (4.9)
300 μm (No. 50)	25		6.14 (30)		1.54 (7.5)
150 μm (No. 100)	18		12.29 (60)		2.21 (10.8)
75 μm (No. 200)	6		32.77 (160)		1.97 (9.6)
			Surface Area =		8.67 m2/kg (42.3 ft2/lb)

*Surface area factor is .41 m2/kg (2 ft2/lb) for any material retained above the 4.75mm (No. 4) sieve.

6.06 C.K.E. PROCEDURE   FOR FINE AGGREGATE — The centrifuge kerosene equivalent method involves these steps:

(a) Place exactly 100 g of dry aggregate [representative of the passing 4.75 mm (No. 4) material being used] in a tared centrifuge cup assembly fitted with a screen and a disk of filter paper.

(b) Place bottom of centrifuge cup in kerosene until the aggregate becomes saturated.

(c) Centrifuge the saturated sample for two minutes at a force of 400 times gravity. (For the suggested centrifuge this force can be developed by turning the handle approximately 45 revolutions per minute.)

(d) Weigh sample after centrifuging and determine the amount of kerosene retained as a percent of the dry aggregate weight; this value is called the Centrifuge Kerosene Equivalent (C.K.E.). (Note: Duplicate samples are always prepared in order to balance the centrifuge and to check results. The average of the two

C.K.E. values is used unless there is a large discrepancy, in which case the test is rerun.)

(e) If the specific gravity of the aggregate samples is greater than 2.70 or less than 2.60 make a correction to the C.K.E. value using the formula at the bottom of the chart in Figure 6.2.

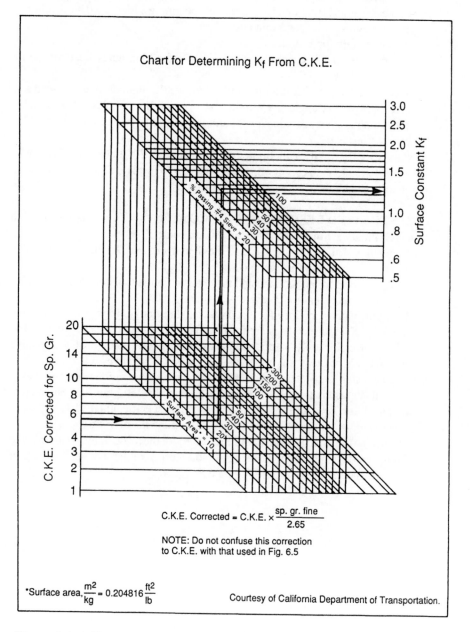

Chart for Determining K_f From C.K.E.

C.K.E. Corrected = C.K.E. $\times \dfrac{\text{sp. gr. fine}}{2.65}$

NOTE: Do not confuse this correction to C.K.E. with that used in Fig. 6.5

*Surface area, $\dfrac{m^2}{kg} = 0.204816 \dfrac{ft^2}{lb}$

Courtesy of California Department of Transportation.

Figure 6.2 – **Chart for determining surface constant for fine material, K_f from C.K.E., Hveem method of design**

6.07 SURFACE CAPACITY TEST FOR COARSE AGGREGATE — The surface capacity (or "oil soak") test for the larger aggregate involves these steps:

(a) Place exactly 100 g of dry aggregate which passes the 9.5 mm (3/8-in.) sieve and is retained on the 4.75 mm (No. 4) sieve into a metal funnel (this fraction is considered to be representative of the coarse aggregate in the mix).

(b) Immerse sample and funnel into a beaker containing SAE No. 10 lubricating oil at room temperature for 5 minutes.

(c) Allow to drain for 2 minutes.

(d) Remove funnel and sample from oil and drain for 15 minutes at a temperature of 60°C (140°F).

(e) Weigh the sample after draining and determine the amount of oil retained as a percent of the dry aggregate weight. (Note: Duplicate samples are prepared to check results. Average value is used unless there is a large discrepancy, in which case the test is rerun.)

(f) If the specific gravity of the aggregate is greater than 2.70 or less than 2.60 make a correction to the percent oil retained using the formula at the bottom of the chart in Figure 6.3.

6.08 ESTIMATED DESIGN ASPHALT CONTENT — These steps are used to make a preliminary estimate of the design asphalt content:

(a) Using the C.K.E. value obtained and the chart in Figure 6.2, determine the value K_f (surface constant for fine material).

(b) Using the percent oil retained and the chart in Figure 6.3, determine the value K_c (surface constant for coarse material).

(c) Using the values obtained for K_f and K_c and the chart in Figure 6.4, determine the value K_m (surface constant for fine and coarse aggregate combined). $K_m = K_f +$ correction to K_f. The correction to K_f obtained from Figure 6.4 is positive if (K_c-K_f) is positive and is negative if (K_c-K_f) is negative.

(d) The next step is to determine the approximate bitumen ratio for the mix based on cutback asphalts of RC-250, MC-250 and SC-250 grades. With values obtained for K_m, surface area, and average specific gravity, use Case 2 procedures of the chart in Figure 6.5 to determine the oil ratio.

(e) Determine the asphalt content (bitumen ratio) for the mix (Figure 6.6) corrected for the grade to be employed, using the surface area of the sample, the grade of asphalt, and the oil ratio from Figure 6.5.

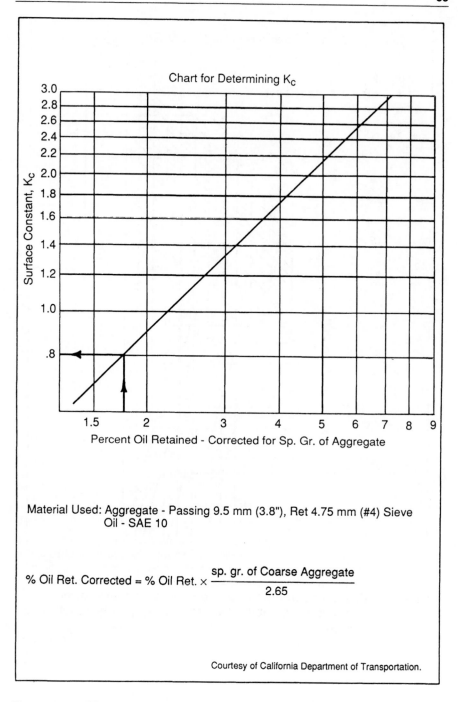

Figure 6.3 – **Chart for determining surface constant for coarse material, K_c, from coarse aggregate absorption, Hveem method of design**

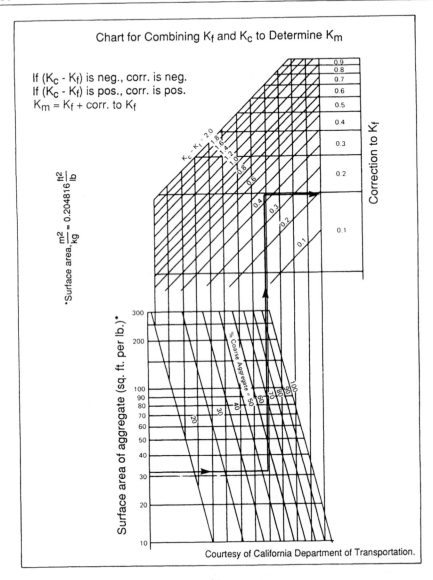

Figure 6.4 – **Chart for combining K_f and K_c to determine surface constant for combined aggregate, K_m, Hveem method of design**

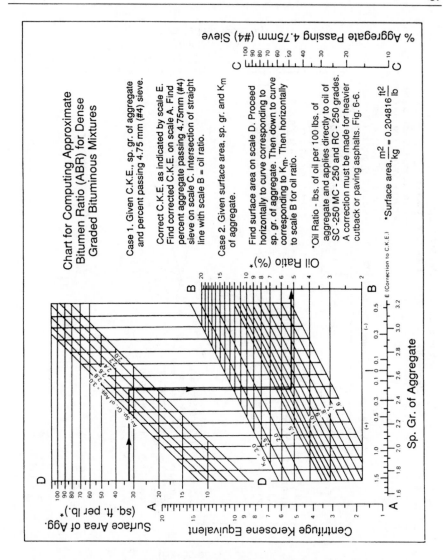

Figure 6.5 – **Chart for computing oil ratio for dense-graded bituminous mixtures, Hveem method of design (Chart courtesy of California Department of Transportation)**

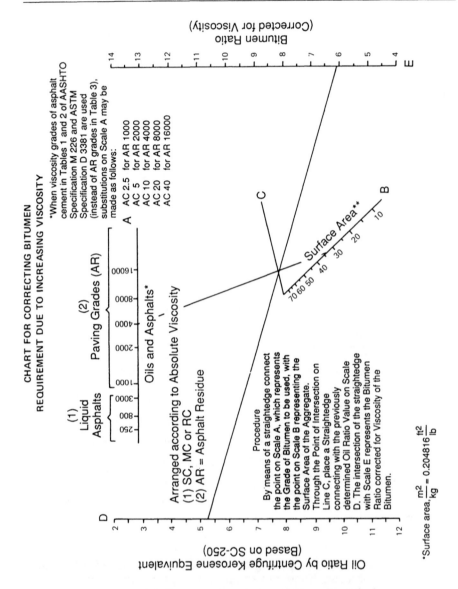

Figure 6.6 – **Chart for correcting bitumen requirement due to increasing viscosity of asphalt, Hveem method of design (Chart courtesy of California Department of Transportation)**

EXAMPLE

To demonstrate the use of the charts in Figures 6.2 through 6.6, assume that these conditions apply to a paving mix using AC-10 viscosity grade asphalt cement:

Specific Gravity, coarse (bulk)[1]   = 2.45
Specific Gravity, fine (apparent)[2] = 2.64
Percent Passing No. 4   = 45

$$\text{Avg. Sp. Gr.}[3] = \frac{100}{\frac{55}{2.45} + \frac{45}{2.64}} = 2.53$$

Surface Area of Aggregate Grading = 6.6 m^2/kg (32.4 ft^2/lb)
C.K.E.   = 5.6
Percent Oil Retained, coarse   = 1.9
(corrected for specific gravity, this value is 1.7 percent. See Figure 6.3)

From Figure 6.2 determine K_f as 1.25.
From Figure 6.3 determine K_c as 0.8.
From Figure 6.4 determine K_m as 1.15.
From Figure 6.5 determine the oil ratio for liquid asphalt as 5.2 percent.
From Figure 6.6 determine design asphalt content (bitumen ratio) for AC-10 asphalt as 6.1 percent by weight of dry aggregate.

Notes:
1. California DOT Manual of Tests, No. 206, which is a modification of AASHTO T85 (ASTM C127).
2. California DOT Manual of Tests, No. 208, which is a modification of AASHTO T133 (ASTM C188).
3. As defined by California DOT Manual of Tests, No. 303.

C. Preparation of Test Specimens

6.09 GENERAL — In designing a paving mix by the Hveem method, a series of stabilometer test specimens is prepared for a range of asphalt contents both above and below the approximate design asphalt content indicated by the CKE procedure.

For hot-mix designs using an average aggregate, tests should be scheduled by preparing one specimen with the amount of asphalt as determined by the CKE procedure, two above the CKE amount in 0.5 increments, and one 0.5 percent below the CKE amount (total of four specimens, each with a different asphalt content). For mixes thought to be critical (i.e. sensitive to asphalt content), the steps in asphalt content are lowered to 0.3 percent and tests are scheduled for the approximate asphalt content indicated by the CKE procedure; three above the CKE amount in 0.3 percent increments, and one 0.3 percent below the CKE amount. For highly absorptive aggregates and non-critical mixes, increase the steps in asphalt content to 1.0 percent and use more specimens as necessary. Regardless of these general rules for preparing stabilometer specimens, the series of test specimens should have at least one specimen containing an excess of asphalt as indicated by moderate or heavy flushing after compaction.

In addition, two swell test specimens are later prepared at the same design asphalt content as determined from tests on the series of specimens prepared for stabilometer tests.

Therefore, for a normal mix design study a total of six test specimens will usually be required. Although each test specimen will normally require only 1.2 kg of aggregate, the minimum aggregate requirements for a series of test specimens should be at least 18 kilograms (40 lb.) to provide for additional tests that may be required.

6.10 EQUIPMENT — The equipment required for the preparation of test specimens is:

(a) *Pans*, 250 mm (10 in.) diameter x 50 mm (2 in.) deep, for quartering and mixing fine aggregate.
(b) *Pans*, 200 mm (8 in.) diameter x 45 mm (1-3/4 in.) deep, for batching and heating aggregates.
(c) *Pans*, 305 mm (12 in.) diameter x 64 mm (2-1/2 in.) deep, for mixing aggregate and asphalt.
(d) *Pans*, 280 mm (11 in.) x 180 mm (7 in.) x 25 mm (1 in.) for curing mix.
(e) Large *sample splitter* for mixing and quartering fine aggregate.
(f) *Electric hot plate*, with a surface measuring at least 460 mm (18 in.) x 305 mm (12 in.), for heating aggregates, asphalt, and equipment as required.
(g) Large *oven*, thermostatically-controlled, capable of 110°C (230°F) temperature.
(h) Large *oven*, thermostatically-controlled, capable of 60°C (140°F) temperature.
(i) Large *oven* for drying and preheating aggregates, capable of temperatures up to 165°C (325°F).
(j) Large *scoop* for handling hot aggregates.
(k) *Beakers*, 800 ml, for adding asphalt.
(l) *Thermometer*, armored, 35°C (100°F) to 205°C (400°F).
(m)*Balance*, minimum 5 kg capacity, sensitive to 0.1 g for weighing aggregates and asphalt.
(n) Small pointed *mixing trowel*.
(o) Large *mixing spoon*.
(p) *Mechanical mixer* (optional).
(q) *Mechanical Compactor* (see Figure 6.7) designed to consolidate the material by a series of individual "kneading action" impressions made by a roving ram having a face shaped as a sector of a 101.6 mm (4 in.) diameter circle (see Figure 6.8). The compactor must be capable of exerting a force of 34.5 kPa (500 psi) beneath the tamper foot. Accessories with the compactor should include two mold holders, an insulated feeder trough 460 mm (18 in.) long x 102 mm (4 in.) wide x 64 mm (2 1/2 in.) deep, a paddle shaped to fit the trough, and a round-nosed steel rod 9.5 mm (3/8 in.) diameter x 406 mm (16 in.) long.

Figure 6.7 – **Mechanical kneading compactor for the preparation of Hveem test specimens**

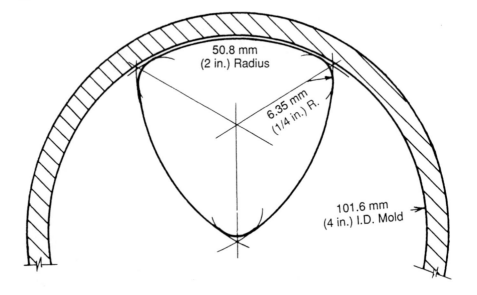

Figure 6.8 – **Diagram of tamping foot for mechanical kneading compactor**

(r) Steel *compaction molds,* 101.6 mm (4 in.) inside diameter x 127 mm (5 in.) high x 6.4 mm (1/4 in.) wall thickness.

(s) *Paper Disks,* heavy paper, 100 mm (4 in.) in diameter, to place in bottom of mold during compaction.

(t) Hydraulic *compression machine,* 222 kN (50,000 lb) capacity.

(u) Steel *shim,* 6.4 mm (1/4 in.) thick x 19.0 mm (3/4 in.) wide x 63.5 mm (2-1/2 in.) long.

(v) *Gloves,* heavy and sturdy, for handling hot equipment.

(Note: See Articles 6.04 and 6.15 for additional equipment requirements.)

6.11 BATCH WEIGHTS — These guidelines are suggested for estimating the aggregate requirements:

(a) Compute batch weights for the blend and gradation of aggregates desired. Suggested procedures for computing batch weights are presented in Chapter 3.

(b) The necessary dry weight of the aggregate for the stabilometer specimens is that which will produce a compacted specimen 63.5 +/- 1.3 mm (2.5 +/- 0.05 in.) in height. This volume of aggregate will normally weigh about 1200 grams. To determine the exact batch weight, it is generally desirable to prepare a trial specimen prior to preparing the actual aggregate batches. If the trial specimen height falls outside the limits, the amount of aggregate used for the specimen may be adjusted using:

For International System of Units (SI),

$$\text{Adjusted mass of aggregate} = \frac{63.5 \text{ (mass of aggregate used)}}{\text{Specimen height (mm) obtained}}$$

For U.S. Customary Units,

$$\text{Adjusted weight of aggregate} = \frac{2.5 \text{ (weight of aggregate used)}}{\text{Specimen height (in.) obtained}}$$

6.12 PREPARATION OF BATCH MIXES — These steps are provided as a guide in preparing the mixtures for testing:

(a) Weigh the various-sized fractions of dry aggregates into suitable pans in accordance with the calculated batch weights. (See Chapter 3)

(b) Thoroughly mix each individual batch of aggregate and preheat in oven to desired mixing temperature. Asphalt should be preheated at the same time. The temperature of the aggregate and the asphalt at the time mixing begins is indicated below for the paving grade of asphalt cement being used.

	Temperature Range	
Grade	Minimum	Maximum
AC-2.5, AR-1000, or 200-300 Pen.	99°C (210°F)	121°C (250°F)
AC-5, AR-2000, or 120-150 Pen.	110°C (230°F)	135°C (275°F)
AC-10, AR-4000, or 85-100 Pen.	121°C (250°F)	149°C (300°F)
AC-20, AR-8000, or 60-70 Pen.	132°C (270°F)	163°C (325°F)
AC-40, AR-1600, or 40-50 Pen.	132°C (270°F)	163°C (325°F)

(c) When the aggregates and asphalt have reached the desired mixing temperature, form a crater in the aggregates and weigh in asphalt in accordance with the calculated batch weights.

(d) Place pan containing aggregates and asphalt for batch mix on hot plate to maintain mixing temperature. Vigorously mix aggregates and asphalt by hand with a pointed trowel or by mechanical mixing until all particles are coated. Take special precaution not to overheat the materials.

(e) After mixing is complete, transfer the batch mix to a suitable flat pan and cure for 2 to 3 hours at a temperature of 146 ± 3°C (295 ± 5°F) in an oven equipped with forced draft air circulation. (Another procedure that has been used is to cure the mix for 15 to 18 hours at 60°C ± 2.8°C (140°F ± 5°F).)

(f) After curing is complete, place batch mix in heating oven and reheat mixture to 110°C (230°F). The batch mix is then ready for compaction.

6.13 COMPACTION — The compaction of the test specimen is accomplished by means of the mechanical compactor that imparts a kneading action type of consolidation by a series of individual impressions made with a ram having a face shaped as a sector of a 101.6 mm (4 in.) diameter circle (Figure 6.8). With each push of the ram, a pressure of 3.45 MPa (500 psi) is applied, subjecting the specimen to a kneading compression over an area of approximately 2000 mm^2 (3.1 in.^{2}). Each pressure is maintained for approximately 2/5ths of a second. The detailed compaction procedure is:

For Stabilometer Specimens:

(a) Preheat the compaction molds, feeder trough and round-nosed steel rod to approximately the mix compaction temperature.

(b) Heat the compactor foot to a temperature that will prevent the mix from adhering to it. The temperature of the compactor foot may be controlled by a variable transformer.

(c) Place the compaction mold in the mold holder and insert a 100 mm (4 in.) diameter paper disk to cover the base plate. So the base plate will act as a free-fitting plunger during the compaction operation, the steel shim is temporarily placed under the edge of the mold.

(d) Spread the prepared mixture uniformly on the preheated feeder trough. Using a paddle that fits the shape of the trough, transfer approximately one-half of the mixture to the compaction mold (See Figure 6.9).

(e) Rod the portion of the mix in the mold 20 times in the center of the mass and 20 times around the edge with the round-nosed steel rod (See Figure 6.10). Transfer the remainder of the sample to the mold and repeat the rodding procedure.

Figure 6.9 – **Transfer of mix to mold**   Figure 6.10 – **Rodding mix in mold**

(f) Place mold assembly into position on the mechanical compactor and apply approximately 20 tamping blows at 1.7 MPa (250 psi) pressure to achieve a semi-compacted condition of the mix so that it will not be unduly disturbed when the full load is applied. The exact number of blows to accomplish the semi-compaction shall be determined by observation. The actual number of blows may vary between 10 and 50, depending upon the type of material, and it may not be possible to accomplish the compaction in the mechanical compactor because of undue movement of the mixture under the compactor foot. In these instances use a 178 kN (40,000 lb.) static load applied over the total specimen surface by the double plunger method, in which a free-fitting plunger is placed below and on top of the sample. Apply the load at the rate of 1.3 mm (0.05 in.) per minute and hold for 30 ± 5 seconds.

(g) After the semi-compaction, remove shim and release mold tightening screw sufficiently to allow free up-and-down movement of mold and about 3 mm (1/8 in.) side movement of mold.

(h) To complete compaction in the mechanical compactor, increase compactor foot pressure to 3.45 MPa (500 psi) and apply 150 tamping blows.

(i) Placed the mold and specimen in an oven at 60°C (140°F) for 1 hour, after which a "leveling-off" load of 56 kN (12,600 lb.) is applied by the "double-plunger" method [head speed = 6 mm/min (0.25 in./min)] and released immediately. (Note: The specimen shall not be pushed to the opposite end of the mold.)

For Swell Test Specimens:

(a) Prepare the compaction mold by placing a paraffin-impregnated strip of ordinary wrapping paper 19 mm (3/4 in.) wide, around the inside of the mold 13 mm (1/2 in.) to 19 mm (3/4 in.) from the bottom to prevent water from escaping from between the specimen and the mold during the water immersion period of the test. The paper strip is dipped in melted paraffin and applied while hot. Compaction molds are not preheated for swell test specimens.

(b) The remainder of the compaction procedure for swell test specimens is the same as for the stabilometer test specimens except for:

When compaction is completed in the mechanical compactor, remove mold and specimen from compactor, invert mold and push specimen to the opposite end of mold. Apply a 56 kN (12,600 lb.) static load [head speed 6 mm/min (0.25 in/ min)] with the "original" top surface supported on the lower platen of the testing press. It is advisable to place a piece of heavy paper under the specimen to prevent damage to this lower platen.

D. Test Procedures

6.14 GENERAL — In the Hveem method the compacted test specimens are used in these tests and analyses and are normally performed in the order listed:

(a) Stabilometer Test
(b) Bulk Density Determination
(c) Swell Test.

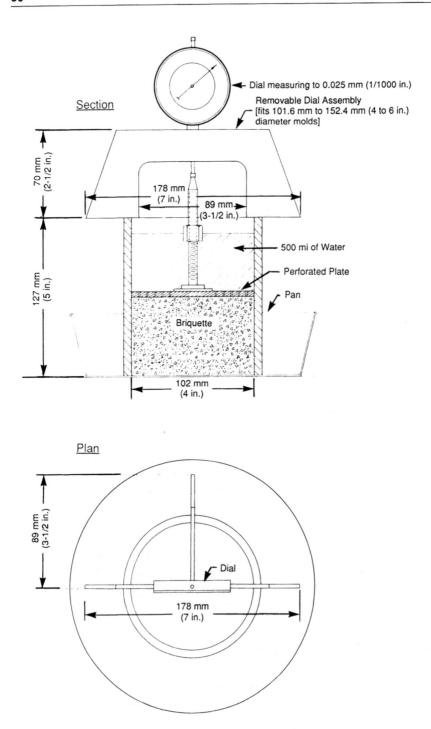

Figure 6.11 – **Swell test apparatus**

The swell test is performed only on specimens prepared for this purpose; the stabilometer and bulk density tests are performed on each of all other test specimens. Figure 6.15 shows a suggested test report form for recording test data and results.

6.15 EQUIPMENT — The equipment required for the testing of the 101.6 mm (4 in.) diameter specimens is:

    (a) *Bronze Disks,* perforated, 98.4 mm (3-7/8 in.) diameter x 3.2 mm (1/8 in.) thick, with adjustable stem, for swell measurement (See Figure 6.11).

    (b) *Dial Gauge,* mounted on tripod, with reading accuracy to 0.025 mm (0.001 in.) (See Figure 6.11).

    (c) *Scale,* graduated to read the volumetric contents of a 101.6 mm (4 in.) inside-diameter mold at 25 ml intervals, for measuring percolation of water during swell test.

    (d) *Aluminum pans,* 190 mm (7-1/2 in.) diameter x 64 mm (2-1/2 in.) deep.

    (e) *Hveem Stabilometer* (see Figures 6.12 and 6.13), complete with accessories including adjustable base, assembly tool, steel follower, and rubber bulb for introducing air into the system.

    (f) *Scale* or other measuring device to accurately determine the height of the compacted test specimen.

6.16 STABILOMETER TEST — These are the steps for measuring the Hveem Stability (refer to Figures 6.12 and 6.13):

    (a) Place specimens for stabilometer tests (compacted and contained in mold) in oven at 60 ± 3°C (140 ± 5°F) for 3 to 4 hours.

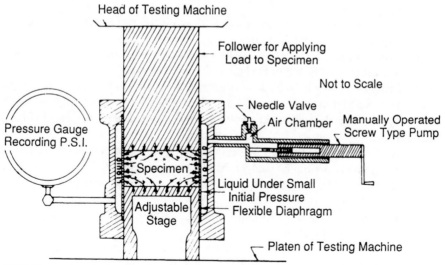

NOTE: The specimen is given lateral support by the flexible sidewall, which transmits horizontal pressure to the liquid. The magnitude of the pressure can be read on the gauge.

Figure 6.12 – **Diagrammatic sketch showing principal features of Hveem stabilometer**

Figure 6.13 – **Hveem
stabilometer**

(b) Adjust compression machine for a head speed of 1.3 mm/min (0.05 in/min) with no load applied.

(c) Check displacement of stabilometer with a calibration cylinder and if necessary adjust to read 2.00 ± 0.05 turns (see Article 6.21).

(d) Adjust the stabilometer base so that the distance from the bottom of the upper tapered ring to the top of the base is 89 mm (3.5 in.).

(e) Every effort should be made to fabricate test specimens with an overall height between 61 mm (2.40 in.) and 66 mm (2.60 in.); however, if the height is outside of this range the stabilometer value should be corrected as indicated in Figure 6.14.

(f) Remove the mold with its specimen from the oven and place on top of stabilometer. Using the plunger, hand lever, and fulcrum, force the specimen from the mold into the stabilometer. Take care that the specimen goes in straight and is firmly seated.

(g) Place follower on top of specimen and position the entire assembly in compression machine for testing.

(h) Using a displacement pump, raise the pressure in the stabilometer system until the test gauge (horizontal pressure) reads exactly 34.5 kPa (5 psi). (Tap test gauge lightly to assure an accurate reading.)

(i) Close displacement pump valve, taking care not to disturb the 34.5 kPa (5 psi) initial pressure. (This step is omitted on stabilometers that are not provided with the displacement pump valve.)

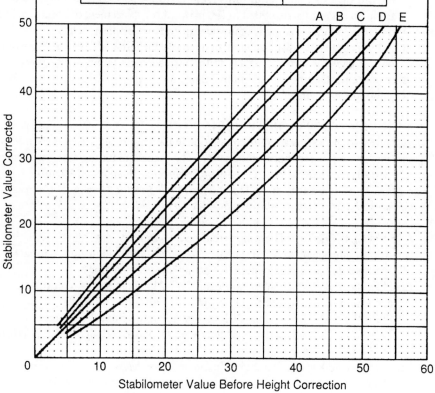

Figure 6.14 – **Chart for correcting stabilometer values to effective specimen height of 64mm (2.50 inches)**

(j) Apply test loads with compression machine using a head speed of 1.3 mm/min (0.05 in/min). Record readings of stabilometer test gauge at vertical test loads of 13.4, 22.3, and 26.7 kN (3,000, 5,000, and 6,000 lbs.)

(k) Immediately after recording the horizontal pressure reading under maximum vertical load [26.69 kN (6,000 lb.)], reduce total load on specimen to 4.45 kN (1,000 lb.).

(l) Open the displacement pump angle valve and by means of the displacement pump, adjust test gauge to 34.5 kPa (5 psi). (This will result in a reduction in the applied press load which is normal and no compensation is necessary).

(m) Adjust dial gauge on pump to zero by means of small thumbscrew.

(n) Turn displacement pump handle smoothly and rapidly (two turns per second) and to the right (clockwise) until a pressure of 690 kPa (100 psi) is recorded on the test gauge. [During this operation the load registered on the testing press will increase and in some cases exceed the initial 4.45kN (1,000 lb.) load. This change in load is normal and no adjustment or compensation is required.] Record the exact number of turns required to increase the test gauge reading from 34.5 kPa (5 psi) to 690 kPa (100 psi) as the displacement on specimen [2.5 mm (0.1 in.) dial reading is equivalent to one turn displacement].

(o) After recording the displacement, first remove the test load and reduce pressure on test gauge to zero by means of the displacement pump; then reverse the displacement pump an additional three turns and remove specimen from stabilometer chamber.

6.17 BULK DENSITY DETERMINATION — The bulk density test is performed on these specimens after the completion of the stabilometer tests as soon as the specimens have cooled to room temperature. The procedure for this test is presented in ASTM D 1188, *Bulk Specific Gravity of Compacted Bituminous Mixtures Using Paraffin-Coated Specimens* or ASTM D 2726, *Bulk Specific Gravity of Compacted Bituminous Mixtures Using Saturated Surface Dry Specimens.*

6.18 SWELL TEST — These steps outline the swell test procedure:

(a) Allow compacted swell test specimen to stand at room temperature for at least one hour. (This is done to permit rebound after compaction).

(b) Place the mold and specimen in 190 mm (7-1/2 in.) diameter x 64 mm (2-1/2 in.) deep aluminum pan (See Figure 6.11).

(c) Place the perforated bronze disk on specimen, position the tripod with dial gauge on mold, and set the adjustable stem to give a reading of 2.54 mm (0.10 in.) on the dial gauge (See Figure 6.11).

(d) Introduce 500 ml of water into the mold on top of the specimen and the measure distance from the top of the mold to the water surface with the graduated scale.

(e) After 24 hours, read the dial gauge to the nearest 0.025 mm (0.001 in.) and record the change as swell. Also, measure the distance from the top of the mold to the water surface with the graduated scale and record the change as permeability or the amount of water in ml that percolates into and/or through the test specimen.

E. Interpretation of Test Data

6.19 CALCULATIONS — There are no calculations required for the swell test since the results are reported directly as differences. The remainder of the calculations are:

(a) *Stabilometer Value.* Calculate as:

$$S = \frac{22.2}{\dfrac{P_h D}{P_v - P_h} + 0.222}$$

where, S = stabilometer value
D = displacement on specimen
P_v = vertical pressure [typically 2.76 MPa (400 psi) = 22.24 kN (5000 lb.) total load]
P_h = horizontal pressure equal to stabilometer pressure gauge reading taken at the instant P_v is 2.76 MPa (400 psi) [22.24 kN (5000 lb.)] total load

(b) *Density and Voids Analysis.* Using the bulk specific gravity of the test specimens and the maximum specific gravity of the paving mixture determined using ASTM D 2041, compute the percent air voids as illustrated in Figure 6.15 and more fully described in Chapter 4.

These values may be plotted as a function of asphalt content as shown in Figure 6.16, similar to the Marshall procedure, to assist in design.

Trial Mix Series: 1-B	**Hot-Mix Design Data**	Project: FI-008-8(3)
53% CA; 47% FA	**by the**	Location: Rye - South
	Hveem Method	

Sp. Gr. Asp. Cem. 1.012 Asp. Cem. AC-10　　　Lab. No. for Asp. Cem. Used: 53-0741
Avg. Bulk Sp. Gr. Agg. = 2.760　　　　　　　　Lab. Nos. for Agg. Used: 53-1252; 53-1253

Gradation, CKE, and Percent Asphalt

Sieve Size	37.5	25.0	19.0	12.5	9.5	4.75	2.36	1.18	600	300	150	75
	1-1/2	1	3/4	1/2	3/8	4	8	16	30	50	100	200
Specification Limits				100	100	90	70	50	29		16	10
					80	70	50	35	18		8	4
% Passing				100	91	76	60	42	32	23	16	12	6
S. A. Factors							.41	.82	1.64	2.87	6.14	12.29	32.77
Surface Area, m2/kg*						.41	.25	.34	.52	.66	.98	1.47	1.97

C.K.E.: FA = 2.8; CA = 2.8 Kf = 1.0; Kc = 1.3; Km = 1.0; Total SA 6.6 m2/kg (32.3 ft2/lb)
Estimated % Asp. Cem. by Wgt. of Agg. using CKE Tests only　　–5.5
Recommended % Asp. Cem. by Wgt. of Agg. using Mix Design Criteria　–5.0

Specimen Identification	A	B	C	D
% Asp. Cem. by Wgt. of Agg.	5.0	5.5	6.0	6.5
% Asp. Cem. by Wgt. of Mix	4.8	5.2	5.7	6.1
Wgt. in Air-grams	1211.0	1223.3	1230.8	1235.9
Wgt. in Water-grams	714.9	723.8	727.6	733.3
Bulk Volume-cc.	496.1	499.5	503.2	502.6
Bulk Sp. Gr.	2.441	2.449	2.446	2.459
Max. Sp. Gr.	2.559	2.540	2.522	2.504
% Air Voids	4.6	3.6	3.0	1.8
Unit Wgt. - Mg/m3	2.439	2.448	2.446	2.457
(lb/ft3)	(152.3)	(152.8)	(152.6)	(153.4)

Total	(lbs)	Unit	(psi)	Stabilometer			
Load kN		Load-Mpa					
2.22	(500)	0.28	(40)	9	9	9	10
4.45	(1000)	0.55	(80)	12	12	15	16
8.90	(2000)	1.10	(160)	15	16	24	26
13.34	(3000)	1.65	(240)	21	22	30	38
17.79	(4000)	2.21	(320)	28	30	42	55
12.24	(5000)	2.76	(400)	36	39	55	83
26.69	(6000)	3.31	(480)	50	52	62	105
Displacement-turns				2.40	2.50	2.46	2.50
Stability Value				48	45	36	25

Cohesiometer

Temperature - °C (°F)				
Effective Height - mm (in.)				
Shot Weight - grams				
Cohesiometer Value				

Jones
　　　　　　　　　　　Inspector

*Surface Area ft2/lb = m2/kg ÷ 0.204816

Figure 6.15 – **Suggested test report form showing test data for a typical mix design by the Hveem method**

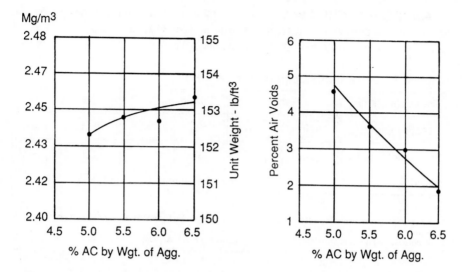

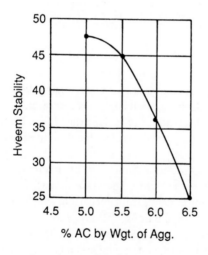

Figure 6.16 – **Test property curves for hot mix design data by the Hveem method**

6.20 DESIGN CRITERIA — The suitability of the hot-mix design by the traditional Hveem method is determined on the basis of whether the asphalt content and aggregate grading will satisfy the requirements in Table 6.2:

Table 6.2 – **Hveem mix design criteria**

Traffic Category	Heavy		Medium		Light	
Test Property	min.	max.	min.	max.	min.	max.
Stabilometer Value	37	–	35	–	30	–
Swell	less than 0.762 mm (0.030 in.)					

NOTES:
1. Although not a routine part of this design method, an effort is made to provide a minimum percent of air voids of approximately 4 percent.
2. All criteria, and not stability value alone, must be considered in designing an asphalt paving mix.
3. Hot-mix asphalt bases that do not meet these criteria when tested at 60°C (140°F) are satisfactory if they meet the criteria when tested at 38°C (100°F) and are placed at 100 mm (4 in.) or more below the surface. This recommendation applies only to regions having a range of climatic conditions similar to those prevailing throughout most of the United States. A different lower test temperature may be considered in regions having more extreme climatic conditions.
4. Traffic Classifications:
   Light: Traffic conditions resulting in a Design EAL <10^4.
   Medium: Traffic conditions resulting in a Design EAL between 10^4 and 10^6.
   Heavy: Traffic conditions resulting in a Design EAL >10^6.

In applying these requirements, the design asphalt content should be the highest percentage the mix will accommodate without reducing stability or void content below minimum values. The design asphalt content is determined from stabilometer values, percent air voids, and observations of surface flushing or bleeding of specimens after compaction. These steps are used to select the design asphalt content:

(a) Using Figure 6.17, insert in Step (1) of the pyramid, the asphalt contents used for preparing the series of mix design specimens. Insert the asphalt contents in order of increasing amounts from left to right with the maximum asphalt content used in the square on the right.

(b) Select from Step (1) the three highest asphalt contents that do not exhibit moderate or heavy surface flushing and record on Step (2). Surface flushing and/or bleeding is considered "Slight" (acceptable) if the surface has only a slight sheen. It is considered "Moderate" (unacceptable) if sufficient free asphalt is apparent to cause paper to stick to the surface but no distortion is noted. Surface flushing is considered "Heavy" (unacceptable) if there is sufficient free asphalt to cause surface puddling or specimen distortion after compaction.

(c) Select from Step (2) the two highest asphalt contents that provide the specified minimum stabilometer value and enter them in Step (3).

(d) Select from Step (3) the highest asphalt content that has at least 4.0 percent air voids and enter in Step (4).

(e) The asphalt content in Step (4) is the design asphalt content. However, if the maximum asphalt content used in the design set (Step 1) is the asphalt content entered on Step (4), additional specimens must be prepared with increased asphalt contents in 0.5 percent increments and a new design asphalt content determination should be made.

Additional mix design voids criteria are contained in Chapter 4 as well as some factors to consider when selecting the final design asphalt content.

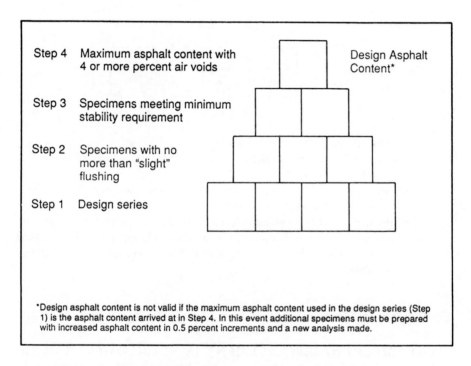

Step 4   Maximum asphalt content with
          4 or more percent air voids   Design Asphalt
                                                                   Content*

Step 3   Specimens meeting minimum
          stability requirement

Step 2   Specimens with no
          more than "slight"
          flushing

Step 1   Design series

*Design asphalt content is not valid if the maximum asphalt content used in the design series (Step 1) is the asphalt content arrived at in Step 4. In this event additional specimens must be prepared with increased asphalt content in 0.5 percent increments and a new analysis made.

Figure 6.17 – **Procedures for selecting design asphalt content, Hveem method of design**

EXAMPLE

This example illustrates determination of the design asphalt content for a Heavy Traffic Category pavement using typical mix design test data shown in Figure 6.15:

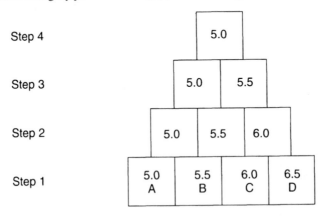

(a) Enter on Step (1) 5.0, 5.5, 6.0 and 6.5.
(b) Specimen D exhibits "Moderate" (unceptable) surface flushing. Enter 5.0, 5.5 and 6.0 on Step (2).
(c) Specimens A and B are the highest asphalt content specimens from Step (2) that provide stabilometer values of 37 or more. Enter 5.0 and 5.5 on Step (3).
(d) Specimen A is the highest asphalt content specimen that has at least 4 percent air voids as shown in Figure 6.16, as desired for heavy traffic. Enter 5.0 on Step (4). The design asphalt content is 5.0 percent using the traditional procedure.

F. Calibration of Hveem Stabilometer

6.21 CHECKING DISPLACEMENT IN STABILOMETER — The stabilometer reading is influenced by mixture displacement in the stabilometer, i.e. the lateral expansion of the specimen during the test. An increase in displacement yields lower P_h readings (or greater apparent stability). For this reason, the displacement must be maintained at a constant value of 2.00 ± 0.05 turns [6.6 cm^3 (0.4 in.^{3})] by using this procedure. Variations in displacement are caused by entrained air and the failure of the diaphragm to fit tightly against the specimen after the initial pressure of 34.5 kPa (5 psi) has been applied. To check the displacement in the stabilometer:

(a) On models with a valve, check displacement pump by closing pump valve and turning the handwheel until tight. If more than one-eighth turn is required to bring the handwheel tight, the pump contains air that must be removed. (Late model stabilometers do not have an angle valve.)

(b) Place stabilometer on adjustable stage and insert a calibration cylinder of steel or brass, 101.6 mm (4.0 in.) in diameter x 139.7 mm (5-1/2 in.) high.

(c) Tighten clamp at base of stabilometer, place stabilometer on compression machine, and apply a confining load of 0.45 kN (100 lb.) to top of the calibration cylinder.

(d) Open angle valve, turn displacement pump handle to right and force fluid into the cell until the stabilometer test gauge registers a horizontal pressure of 34.5 kPa (5 psi).

(e) Set the displacement pump dial gauge to zero with adjustment screw, and turn the displacement pump handle to the right (clockwise) at the rate of two turns per second until pressure on stabilometer test gauge reads 690 kPa (100 psi).

(f) Record displacement as the number of pump handle revolutions required to increase the stabilometer test gauge from 34.5 kPa (5 psi) to 690 kPa (100 psi). [2.5 mm (0.1 in.) dial reading is equivalent to one revolution.]

(g) The stabilometer test gauge should remain stationary at 690 kPa (100 psi). If it visibly decreases, there is a leak that must be located and sealed.

(h) Displacement volume in cm^3 (in^3) may be determined by multiplying the number of revolutions by 3.3 (0.2).

(i) If the displacement does not check within $\pm$ 0.05 turns of the required 2.00 [6.6 cm^3 (0.4 $in.^3$)] an adjustment is necessary. Air is added to the system to increase the displacement or removed from the system to decrease the displacement.

(j) To adjust displacement to the required value, turn pump handle clockwise (to the right) until a pressure of 690 kPa (100 psi) is obtained on test gauge, then turn pump handle counterclockwise (to the left) two turns. Either add or remove air until a pressure of 34.5 kPa (5 psi) is indicated on test gauge. Air may be added to air cell by means of the rubber bulb provided for this purpose. Continue check and adjustment until the required value of displacement is obtained.

(k) Release pressure and remove the calibration cylinder.

6.22 FILLING STABILOMETER WITH LIQUID — If the displacement pump is a rubber gasket "O" ring type, it is recommended that the stabilometer be filled with a machine oil having a Saybolt-Furol viscosity of approximately sixty seconds at 38°C (100°F). If the displacement pump is a packing gland type, the stabilometer should be filled with a solution of 50 percent glycerine and 50 percent distilled water.

It is very important that all air, even the smallest bubbles, be removed from the system. On models equipped with an angle valve, the displacement pump should be filled after closing the angle valve connecting the inner chamber. Remove the small plug in the center of the displacement pump handle, and screw the handle out about 76 mm (3 in.). Fill with liquid and remove all the air, then replace the plug and loosen the packing gland nut. Screw the displacement pump until liquid oozes out around packing gland nut. Next, remove the small plug in the angle valve and, with the valve closed, screw the pump in until all air is removed from angle valve. Remove the filling plug opposite the test gauge and, with the stabilometer resting on the face of the testing gauge, fill with liquid. The rubber diaphragm should be tapped lightly with fingers to remove any air. The machine should be rocked about to eliminate bubbles near the opening.

The above procedure should also be performed on stabilometers that have had the displacement pump converted from a packing gland to an "O" ring type, using the oil described above as the liquid medium.

Chapter 7

Durability of Asphalt Mixtures

7.01 GENERAL — On occasion, a properly-designed asphalt concrete mixture may not perform as expected. The reasons for this lack of good performance may be related to material durability or compatibility. Durability tests may be divided into two categories: aggregates and mixtures. Standard durability tests for the physical properties of the aggregate can be used to evaluate the integrity or quality of the individual particles.

Once the aggregate is acceptable based on these tests, a mixture is then designed in accordance with procedures such as outlined in Chapters 5 or 6. The other category of durability tests is concerned with how that aggregate reacts with the asphalt and how the properties of the finished mix design react in the presence of water.

7.02 AGGREGATE SIZE, SHAPE, AND DURABILITY — Since aggregates typically constitute about 95 percent by weight of the asphalt concrete mixture and supply nearly all of the pavement load bearing capacity, their quality and physical properties are critical to pavement performance. When marginal but locally-available aggregates are used in a mix, the initial savings in construction costs may be completely offset by the loss in durability. Depending on the size of the paving project or the expected traffic loading on the pavement, there are a number of aggregate properties which should be considered.

For good performance, aggregates should be clean and durable. They should also be free of clay or any other deleterious material (ASTM C 142). If significant aggregate breakdown occurs during or after construction, an entirely different mix will exist in the pavement than what was designed in the laboratory. The aggregate gradation, void properties, stability, and asphalt requirements could be significantly altered.

Some of the tests that can be performed to evaluate aggregates being considered for asphalt paving mixes are:

Sulfate Soundness Tests (AASHTO T 104; ASTM C 88) determine the durability or weathering resistance of aggregates by exposing them to a number of soaking and drying cycles in either a sodium or magnesium sulfate solution. The measured amount of aggregate breakdown is typically limited to 12 to 20 percent.

Particle shape (ASTM D 4791), roughness, or angularity is improved by crushing which induces more aggregate interlocking and provides more bonding surface for the asphalt. Depending on the project conditions (i.e. traffic volume), these guidelines are recommended: 60 to 100 percent of the coarse aggregate with one crushed face and possibly 50 to 80 percent with two crushed faces.

Frictional properties of fine aggregate are important. Generally, natural sands are round and smooth and therefore, have lower frictional properties than manufactured sands. For this reason, the percent by weight of total aggregate of the natural sand in a mix is typically limited to 15 to 25 percent in high traffic areas.

Sand Equivalent Test (AASHTO T 176; ASTM D 2419) is used to indicate the amount of "clay-like" material in the fine aggregate. Although the precision of this test is debatable, samples with values below 45 should be further evaluated. The Plasticity Index (AASHTO T 89 and T 90; ASTM D 4318) can also be used to evaluate the plasticity of the fine material.

Los Angeles Abrasion (AASHTO T 96; ASTM C 131) or wear tests can be used to indicate the expected aggregate breakdown during mixing and compaction. For high traffic pavement projects, this measured maximum loss is typically limited to about 40 or 45 percent.

Mix durability has also been related to the amount of fine "dust or dirt" particles in the mixture. Excessive fines can lower the quality of the asphalt film on the aggregate. Depending on the size of these particles, the mix may be stiffer or more tender. The Federal Highway Administration has recommended that mixes, especially in high traffic locations, be designed with *dust-to-asphalt* ratios between 0.6 and 1.2.

7.03 AGGREGATE/ASPHALT COMPATIBILITY — The property of adhesion between asphalt and aggregates in asphalt concrete is very complex and not clearly understood. The loss of bond (stripping) due to the presence of moisture between the asphalt and the aggregate is a problem in some areas of the country and can be severe in some cases. Research has identified five different mechanisms by which stripping may occur. These mechanisms are detachment, displacement, spontaneous emulsification, pore pressure, and hydraulic scouring. These may act individually or together to cause an adhesion failure.

The stripping behavior is complicated by many factors such as type and use of the asphalt-aggregate mix, asphalt characteristics, aggregate characteristics, environment, traffic, construction practice, drainage, and the use of various anti-strip additives. Hydrophobic ("water-hating") aggregates (such as limestone) that have porous, slightly rough surfaces, and surfaces that are clean, dry, and have been aged for a period of time to acquire an organic contamination will generally provide better stripping resistance.

The capacity for water getting into and draining out of a pavement has also been shown to be a critical factor. Stripped, wet mixtures will be much weaker than dry, non-stripped mixtures. However, the effects may be reversible if the asphalt is not completely washed away from the layer. It has also been shown that stripping can be mitigated by changing the asphalt or aggregate, or by adding hydrated lime or a proven additive based on the results of a laboratory strength test. The compatibility of all the actual mix components needs to be checked as a part of the mix design.

Several laboratory test procedures have been developed in an effort to determine the moisture susceptibility of asphalt paving mixtures and to evaluate the potential of certain additives or mix changes. The easiest tests to perform are generally the ones with the least correlation to field performance. The accuracy of many of these simple tests is limited by the subjective nature of the results. To obtain a more objective evaluation, a number of mechanical-immersion tests have evolved that measure the change in a specified property of the compacted mixture, after various types of conditioning processes. This "conditioning" analysis offers a comparison of the dry mixture properties to those of wet mixtures. The design asphalt content selected using measurements of mixes in dry conditions may show substantially different properties under wet conditions. This may vary with the type of aggregate used in the mixture as well as with the properties of the mixture itself. Some mixtures will be reduced in stability while other mixtures may be completely unaffected by the presence of water.

The procedures discussed in this article should be considered as durability tests rather than mix design procedures. However, when the results of the durability tests are considered, modifications in the recommended job-mix formula may be warranted, depending on the anticipated traffic and environmental conditions.

The Immersion-Compression Test (ASTM D 1075 or AASHTO T 165) uses a ratio of the average unconfined compressive strength of a conditioned specimen to an unconditioned specimen. The conditioned specimens are soaked in a 49°C (120°F) water bath for four days or alternatively, for 24 hours at 60°C (140°F). The Asphalt Institute recommends a minimum ratio of retained strength using the immersion-compression test of 75 percent. (Caution: This procedure may produce misleading results when porous aggregates are involved.)

Similarly, the Marshall-immersion test uses a ratio of Marshall stabilities. The ratio compares the stability of Marshall specimens after soaking in a 60°C (140°F) water bath for 24 hours to the stability of specimens tested in accordance with ASTM D1559.

Ratios of both indirect tensile strength and resilient modulus have also been used as before and after reference measurements. In the Modified Lottman procedure (AASHTO T 283), specimens are subjected to vacuum saturation and/or exposed to one or repeated freeze-thaw cycles. The Root-Tunnicliff method (ASTM D 4867) uses a slightly different conditioning procedure with an optional 15-hour freezing cycle. In these two tests, the void range of the compacted specimens should be controlled between 6 and 8 percent, and the saturation of the specimen controlled between 55 and 80 percent. These tests are recommended by most agencies for providing the most realistic results compared to the field. Tensile strength ratios (with conditioning to without conditioning) of about 70 to 75 percent are suggested minimum values. However, the actual value of the test measurement after conditioning should also be examined with respect to some minimal required value. The ratio could be high and yet the actual values, both with and without conditioning, could be too low for adequate performance.

The advantage of these mechanical-immersion tests is that the actual mix components can be compacted to a density representative of the field; however, sample preparation is critical. The disadvantage is that most of these procedures are complicated and time consuming. Currently, the accuracy of these conditioning treatments and test methods remains a point of contention. It appears that the actual strength test used may not be as critical as the conditioning procedures used to simulate the field situation.

The boundary line between realistic and overly-severe conditioning, whether it is duration or temperature of soaking, degree of vacuum saturation, or freeze-thaw cycling, is difficult to judge and currently unknown. Optimally, the conditioning stage should be "tied" to the particular environment. Regardless of the test procedure and conditioning method, if the specimen swells more than 2 percent during conditioning, the results should be considered invalid. The test should be rerun at a higher void content.

It has been documented in previous research at the Asphalt Institute and by others that the stripping behavior of various mixes is affected by the amount of asphalt cement surrounding the aggregates and the percentage of air voids through which the moisture must travel. Therefore, it may be practical and desirable to fabricate a few additional specimens to evaluate these additional factors as part of the planned laboratory mix design testing over the entire range of asphalt contents, especially if this possibility is a major concern.

If the mixture fails to meet these test criteria, the designer can modify the mixture in several ways.* Some of those changes are:

1. Increase the asphalt content
2. Use a higher viscosity (heavier) grade of asphalt
3. Provide a cleaner or different aggregate source
4. Add hydrated lime or liquid anti-stripping additive to mix (if benefit is shown in laboratory testing)
5. Possibly blend aggregates to improve gradation and density.

A good deal of judgment is required with these tests; eliminating cheaper local materials based on the results of an overly-demanding test would be extremely unfortunate. In summary, all of the test procedures have been shown occasionally to provide incorrect results when compared to actual performance; therefore, the indications from these tests should not be considered as ultimate proof of compatibility.

---

* For more information on stripping see the Asphalt Institute publication, *Cause and Prevention of Stripping in Asphalt Pavements* (ES-10).

Chapter 8

Field Verification of Mix Design

8.01 GENERAL — Hot mix asphalt mixtures have traditionally been designed in the laboratory, produced and constructed in the field, and then tested and evaluated based on the performance of the pavement after a few years of traffic. Although these practices are clearly related, they are often considered separate activities.

The goal of mix design is to arrive at a starting point, the job mix formula, for establishing process mixing control and uniformity. Field verification of the hot mix asphalt (HMA) involves testing and analyzing the field-produced mixture to ensure that the criteria established for the particular mixture are being met. Significant equipment and material differences exist between the small scale operation of the mixing bowl and an asphalt mixing facility. Field verification of HMA is necessary to measure what differences exist, and what corrective measures need to be taken.

Field verification is one part of a total quality assurance system designed to assure that the quality of the construction and materials conforms with the plans and specifications under which it was produced. Activities that occur under the umbrella of this total system are:

- quality control practices designed to monitor the product manufacturing process and
- acceptance sampling and testing used to assure that satisfactory quality control has been exercised to attain the proper specification compliance.

A sample quality assurance-quality control schedule is given in Table 8.1.

Field verification is a part of the quality control used in asphalt mixture production. A field verification program for HMA is implemented to prevent the production of substandard or out-of-specification material rather than to document the degree of noncompliance.

8.02 FIELD VERIFICATION TESTING — Specific properties of the field-produced mixture are measured and compared to the job-mix formula. The field verification tests that are used will vary depending on the design procedure specified by the controlling agency. Currently, most organizations that incorporate field verification use the Marshall mix design method because the compaction equipment is more portable, making it economical for field verification techniques. Texas gyratory compaction is also well suited. While some test procedures are governed by strict test methods, others have more than one alternative option. The following sections discuss the tests used to verify the job-mix formula.

Table 8.1 – **Sample quality assurance-quality control schedule**

I.	Pre-production Sampling & Testing
	A. Aggregate for mix design
	B. Mineral filler/additives, if necessary
	C. Asphalt material from proposed source ,
II.	Job-Mix Formula Verification
	A. Aggregate gradation
	B. Asphalt content
	C. Air voids, VMA, and voids filled
	D. Marshall properties, where applicable
III.	Mix Properties During Production
	A. Maximum theoretical specific gravity (Rice)
	B. Bulk specific gravity for air voids
	C. Aggregate gradation
	D. Asphalt content
IV.	In-Place Acceptance Testing
	A. Air voids/in-place density
	B. Thickness
	C. Smoothness/Ride quality
	D. Profile

Asphalt Content

Many methods can be used to determine asphalt content. The most frequently-used method to date is the extraction test which separates the asphalt and aggregate using a solvent (AASHTO T164; ASTM D2172). "Automatic recordation" can be used to calculate asphalt content if the asphalt mixing facility makes detailed measurements of the materials used during production. Properly calibrated nuclear asphalt content gauges can provide measurements on the produced mixture (AASHTO T287; ASTM D4125). These gauges increasing in use due to environmental constraints being placed on the chlorinated solvents used in extractions.

Aggregate Gradation

Various ways also exist to determine aggregate gradation. The aggregate belt feed or hot bins are sometimes sampled prior to mixing with asphalt (see Article 3.03). However, extraction testing of the plant-mixed material is the only true measurement for the aggregate gradation of the final mixture.

Maximum Specific Gravity

The theoretical maximum specific gravity, G_{mm}, of the bituminous paving mixture (AASHTO T209; ASTM D2041) is a key measurement during both laboratory mix design and field verification. Multiplying G_{mm} by the unit weight of water, γ_w, will yield the theoretical maximum density of an asphalt mixture. Also called the "Rice" specific gravity after its developer, G_{mm} is the ratio of the weight in air of a unit volume of an uncompacted asphalt cement and aggregate mixture to the weight of an equal volume of water. Using a vacuum procedure to remove entrapped air from the mixture, the test determines the volume of the asphalt mix in a voidless state.

EXAMPLE

Theoretical maximum specific gravity = G_{mm} = 2.438
Unit weight of water = γ_w = 1,000 kg/m^3 (62.4 lb/ft^3)

Maximum density = G_{mm} x γ_w = 2.438 x 1,000 (62.4) = 2,438 kg/m^3 (152.1 lb/ft^3)

The theoretical maximum specific gravity is used to calculate percent air voids and relative density of both laboratory compacted samples and field compacted pavement cores.

Bulk Specific Gravity
A sample of the plant-produced mixture is compacted using the same procedure used in the mix design (such as 75-blow Marshall). The compacted sample is then used to determine the bulk specific gravity, G_{mb}, of the hot-mix asphalt (AASHTO T166; ASTM D1188 or D2726). Multiplying G_{mb} by the unit weight of water will yield the bulk density of the compacted sample.

EXAMPLE

Bulk specific gravity = G_{mb} = 2.344
Unit weight of water = γ_w = 1,000 kg/m^3 (62.4 lb/ft^3)

Bulk density = G_{mb} x γ_w = 2.344 x 1,000 (62.4) = 2,344 kg/m^3 (146.3 lb/ft^3)

Air Voids
Since G_{mb} is measured on the compacted sample, the measurement includes air contained in the sample. The percent air voids, V_a, of the compacted mixture is calculated using the bulk and maximum theoretical specific gravities in this equation:

$$V_a = \frac{(G_{mm}-G_{mb})}{G_{mm}} \times 100$$

EXAMPLE

G_{mm} = 2.438 G_{mb} = 2.344

$$V_a = \frac{(2.438 - 2.344)}{2.438} \times 100 = 3.9 \text{ percent}$$

Stability and Flow

These properties can be measured on the laboratory compacted samples of field-produced material, and some agencies include them in field verification tests. However, their reliability as quality control tests is less than density/voids analyses, since Marshall stability and flow values are affected by many different aggregate and asphalt properties. If the volumetric properties (voids, asphalt content, gradation) of the mixture are properly controlled, then stability and flow will normally meet the appropriate specifications.

8.03 DENSITY SPECIFICATIONS — Field verification of the HMA involves testing and analyzing the field-produced mixture to ensure that the criteria established for the particular mixture are being met. In most cases, density specifications are used to judge the acceptability of compaction during construction.

The goal of compacting an HMA pavement is to achieve an optimum air void content and provide a smooth, uniform mat. The resultant, in-place air voids of the HMA is probably the single most important factor that affects performance of the mixture throughout the life of the pavement.

The activities involved with the proper design, production, placement, and compaction of the asphalt mixture are all united to achieve the in-place density of the HMA pavement—and ultimately determine whether the pavement will perform as expected. The density specifications to which the pavement is built are used to stipulate the amount of compaction achieved.

A "method" specification has no reference density against which the in-place density and air voids are compared. This type of specification contains items such as number and type of rollers to be used, number of passes of each roller, use of temperature measurements, descriptions such as "surface is rolled until free of roller marks," etc. Judgment is the primary decision tool for determining optimum compaction when using this type of specification. Method specifications are generally only applicable for smaller projects with light traffic, or thin lift construction (one inch or less), such as leveling courses and thin HMA overlays. In these cases, cost and the inability to obtain meaningful data from thin, in-place pavements preclude the use of a reference density specification.

A density specification represents a comparison between the in-place density of the pavement that is achieved after final compaction, and a reference density. One of three reference densities are typically used in density specifications:

Laboratory Density

This method compares in-place density to a laboratory-compacted sample of field-produced asphalt mix, and is particularly applicable to Marshall compaction procedures. A reference density is established to determine the ultimate compactability of the mixture end product. The field-produced HMA is compacted using the same compactive effort used during design (e.g. 50 or 75 blows for Marshall compaction) and the laboratory density is measured using the bulk specific gravity test.

In terms of specification compliance, an agency compares the in-place core density to the reference density in the form of a ratio:

$$\text{Percent of Laboratory Density} = \frac{\text{In-Place Density} \times 100}{\text{Laboratory Density}}$$

When it has been verified that the field-produced mix matches the mix design properties, the laboratory compacted samples provide the same air void content as used in the mix design, typically four percent. If an in-place air voids content of 8 percent is desired, the in-place density should be 96 percent of the reference laboratory density.

Maximum Theoretical Density

The maximum theoretical density provides the unit weight of the mix as if it were compacted to no air voids. Using the Rice test method, the maximum theoretical density of the field-produced mixture is determined as the reference density.

The relative density of the in-place pavement is again calculated as the ratio of the in-place density to the maximum theoretical density:

$$\text{Percent of Max. Theor. Density} = \frac{\text{In-Place Density} \times 100}{\text{Max. Theor. Density}}$$

Since the maximum theoretical density represents a voidless mixture, when an in-place air voids content of 8 percent is desired, the in-place density should be 92 percent of the reference maximum theoretical density.

Control Strip Density

This process calls for the construction of a pavement control strip of a minimum length or volume of mix at the start of each lift being laid. (The control strip is part of the paving project.) After compaction is completed, a specified number of bulk specific gravity (density) tests are measured on core samples taken from random locations within the control strip and averaged to obtain the reference density.

The reference control strip density must then be compared to either the laboratory or theoretical maximum density of the field-produced HMA to determine if densification is adequate and accepted. Once an acceptable control strip has been obtained, since the in-place density is exactly the reference density, 100 percent of the reference density is the desired density during construction.

Good performance with any density specification that involves the use of a reference density depends on such factors as:

- properly-designed mixtures, proper sampling and handling procedures of the loose samples from the mixing facility,
- proper field laboratory testing procedures, especially correct compaction techniques,
- proper sampling, handling, and testing of the in-place core pavement samples, and
- adequate field confinement.

The relationship between the reference density measurements and the air voids of the in-place pavement is shown in Figure 8.1. An in-place air voids benchmark equal to 8 percent is depicted against each type of reference density.

It should be noted that while the comparison between maximum theoretical density and in-place air voids content is a consistent one, the relationships between the other two reference density types and in-place air voids will shift up or down depending on the actual mix design and compaction criteria used in the specification. For example, if the mix design air voids is five percent, then 100 percent of laboratory density would be at five percent air voids. If the same compaction criteria of 96 percent of laboratory density were used, this would yield an in-place air void content of nine percent (not 8 percent).

The use of reference density specifications (laboratory compaction, maximum theoretical and control strip) is appropriate for all projects with a lift thickness greater than one inch. Each of the reference density specification procedures have additional

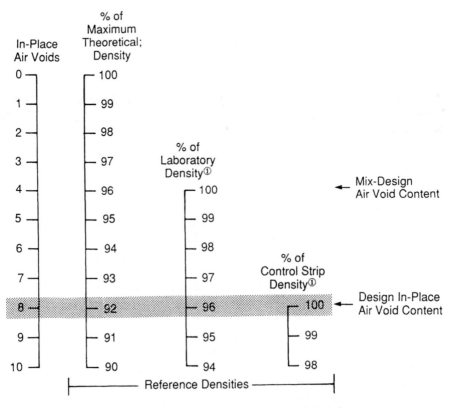

① Relative alignment of these two scales depends on actual mix design and compaction criteria.

Figure 8.1 – **Relationship between the reference density measurements and the air voids** considerations that may make one more favorable than another on a particular project. These considerations include traffic volume, subgrade support, size of the project, construction and testing schedules, and any lift thickness variation.

When a specific reference density procedure (laboratory, maximum theoretical, or control strip) is chosen for a project, this same reference density process should be used throughout the testing and verification of in-place density. This will ensure that a valid comparison exists in the determination of density compliance. A higher degree of compaction monitoring is necessary in the initial stages of the construction process, regardless of which density specification is used, to ensure optimum results from the compaction process.

In addition to minimum compaction, it is also necessary to avoid too much compaction. It has been documented in several cases that even though a mix is designed for four percent air voids, a plant-mix sample containing the proper asphalt content and compacted by the same laboratory method may show much less than the design air voids. Many causes for this occurrence have been proposed:

- moisture may be removed more effectively in the laboratory than in the field
- an excessive amount of baghouse fines may have been added
- highly absorptive aggregates may have been used (field absorption may differ from laboratory absorption)
- poor sampling may have been used to obtain samples for the laboratory design
- materials may have changed between the mix design and field production

If low density (below 3 percent air voids) in a compacted mix is noted, the cause should be determined and corrected. This problem may also require that the mix be adjusted or redesigned.

8.04 DATA ANALYSIS — Field verification involves two different levels of analysis performed on the HMA. The first involves analyzing the mixture on the first day of full production to compare the mixture to the job-mix formula. The second uses day-to-day field verification tests performed to determine if the mixture properties have exceeded production tolerance limits.

Job-Mix Formula Verification
Asphalt content, gradation, and voids analysis tests are performed to compare field-produced mixture properties with the job-mix formula. These tests will indicate if the aggregate characteristics have varied since the mix design, and may indicate if problems exist from possible changes in the aggregate after processing through the dryer.

At this point, the field verification results may show that changes are necessary to meet the job-mix formula. For example, minor changes in the asphalt content may bring a mixture back within the tolerances of the void requirements. Alternatively, if the mixture is meeting overall agency specifications but not the mix design targets, the job-mix formula can be adjusted to accept these new targets. Finally, any dramatic differences between the laboratory design and field-produced mixture may necessitate a new mix design using the actual production materials.

Daily Mix Verification

Once the job-mix formula has been verified, daily testing can provide an early warning by indicating if the mixture properties deviate from the specifications. This daily verification is a part of plant process control that can identify potential problems before many tons of mix have been placed in the field.

Daily field verification tests are typically performed on random samples taken from a set quantity of material called a lot. A lot is typically a day's production or a given tonnage of material. It is important to use random sampling techniques (see *Principles of Construction of Hot-Mix Asphalt Pavements,* MS-22, Asphalt Institute) so that an unbiased evaluation of the material can be made. After the material samples are taken from each lot (a minimum of four is recommended), the verification tests are performed on each sample.

Daily field verification test values are plotted on control charts. Continuous plots of mix data such as percent air voids, asphalt content, and aggregate percentages passing certain sieves such as 4.75mm (No. 4), 600μ m (No. 30), and 75μ m (No. 200) provide a graphic representation of the production process. Target values and tolerance limits are set for each material property, and the production values plotted in relation to these limits can be used to analyze process control.

Figure 8.2 shows a set of control charts of asphalt content during production. The top chart shows the value of each asphalt content test performed. The bottom chart shows the running or moving average of the asphalt content data. The running average is calculated from a subgroup of consecutive test values, typically three to five values per subgroup. After each test is performed, the new test value replaces the oldest test value in the subgroup to calculate the new running average.

When analyzing field verification data, it is important to recognize sources of variation in the data. These sources include variation in the testing and sampling procedures, normal variations in the materials and production process, and variations due to problems in production. Following the testing and sampling procedures exactly as specified will help eliminate this variation. Obviously, adjusting the production process on the basis of erroneous test results is not desirable.

The control charts can help differentiate between variation inherent in the material and production variation and they can provide early signals of problems that need attention. The data should be dispersed randomly about the target value and between the control limits. A few possible indications of existing or upcoming problems are:

- values consistently higher or lower than the average,
- gradual or erratic shifts in the data, and
- systematic cycling of the data.

Based on the present knowledge of plant production and pavement behavior, field verification must be utilized to manage the process of asphalt mixture production to minimize the variability between mix design goals set in the laboratory and actual mix results achieved in the plant.

The in-place air voids of the HMA after compaction is probably the single most important factor that affects performance of the mixture throughout the life of the

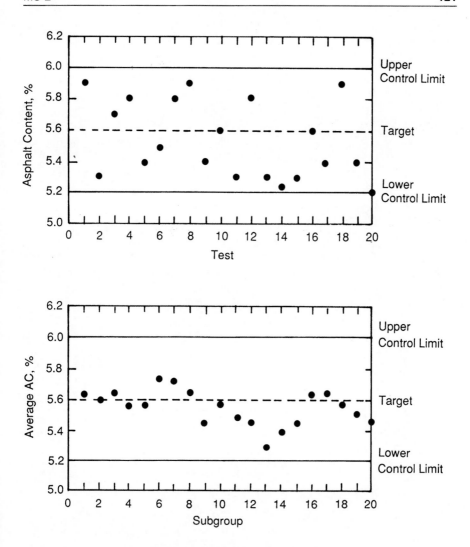

Figure 8.2 – **Typical quality control charts during mixture production**

pavement. However, specifying compaction is not sufficient for ensuring the success of a paving project. Compaction specifications are the final step in the total quality management of the HMA construction process. Proper mix design, production, field verification, and construction procedures must be integrated within the project requirements to achieve a quality product.

Appendix

Mix Design Using RAP

A.01 GENERAL—This appendix presents the step-by step process necessary to incorporate reclaimed asphalt pavement (RAP) and reclaimed aggregate materials (RAM) into asphalt mix design. The steps include proportioning the reclaimed materials, selecting the grade and quantity of asphalt cement (plus recycling agent, if needed) and preparing a final design for the recycled mixture. This is the hot-mix method of recycling, using from 10 to 70 percent reclaimed asphalt pavement. Batch plants can handle up to 50 percent (without some auxiliary method of preheating RAP), with the most practical range being 10 to 35 percent; drum-mix plants can handle up to 70 percent, with 10 to 50 percent being a practical range. Due to the variability of RAP material, experience has generally shown that mixture quality can be more easily controlled by using less than 25 percent RAP. Complete details can be found in the Asphalt Institute's *Asphalt Hot-Mix Recycling* manual (MS-20).

A.02 PREPARATORY STEPS—This mix-design procedure uses either the Marshall or the Hveem method:

The material from a reclaimed asphalt pavement is blended with reclaimed aggregate materials and/or new aggregate that is required to obtain a combined aggregate gradation meeting the specification requirements. Once the relative aggregate proportions are determined, a total asphalt demand is calculated. A grade of new asphalt is then selected (plus recycling agent, if needed) to restore the aged asphalt and provide a final binder that meets the functional requirements of the asphalt specifications while satisfying the asphalt demand of the mix. Following these determinations, the mix design by either the Marshall or Hveem procedure is performed and the exact quantity of total binder determined.

A.03 RAP MIX DESIGN—The composition of the reclaimed materials must be determined. This includes the aggregate gradation, the asphalt content, and the properties of the asphalt cement. Viscosity at 60°C (140°F), (ASTM D 2171; AASHTO T 202), is the test measurement used in this procedure to identify asphalt in the reclaimed asphalt pavement and in the recycled mixture.

Figure A.1 is a flow chart setting forth the steps for this design procedure. The steps are:

(1) *Combined Aggregates in the Recycled Mixture* - Using the gradation of the aggregate from the reclaimed asphalt pavement, the reclaimed aggregate material, if any, and new aggregate, a combined gradation meeting the desired specification requirements is calculated.

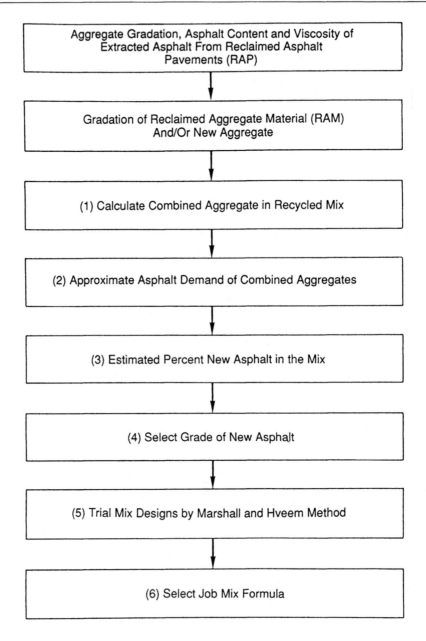

Figure A.1 – **Flow chart for recycling hot mix design procedure**

After the blend of aggregate (aggregate in the RAP, new aggregates and/or RAM) have been established, the amount of new aggregate (and/or RAM) is expressed as r, in percent.

For example, suppose the established blend for a recycled mix was:

> 60% reclaimed aggregate (RAM)
> 15% new aggregate
> 25% RAP aggregate
> 100% Total

The total amount of new aggregate and RAM is 75%. Hence, r=75. Table A.1 contains formulas for proportioning materials for recycled asphalt hot mixes where the blend of aggregates in the mix is kept constant.

(2) *Approximate Asphalt Demand of the Combined Aggregates* -The approximate asphalt demand of the combined aggregates may be determined by the Centrifuge Kerosene Equivalent (CKE) test (See Article 6.03) or calculated by this empirical formula:

$$P = 0.035a + 0.045b + Kc + F$$

where: P = approximate total asphalt demand of recycled mix, percent by weight of mix
   a = percent* of mineral aggregate retained on 2.36mm (No. 8) sieve
   b = percent* of mineral aggregate passing the 2.36mm (No. 8) sieve and retained on the 75μ m (No. 200) sieve
   c = percent of mineral aggregate passing 75μ m (No. 200) sieve
   K = 0.15 for 11-15 percent passing 75μ m (No. 200) sieve
      0.18 for 6-10 percent passing 75μ m (No. 200) sieve
      0.20 for 5 percent or less passing 75μ m (No. 200) sieve
   F = 0 to 2.0 percent. Based on absorption of light or heavy aggregate. In the absence of other data, a value of 0.7 is suggested.

*Expressed as a whole number.

With an approximate asphalt demand established, this will provide a basis for a series of trial mixes for a mix design. Trial mixes will vary in asphalt content in 0.5 increments on either side of the calculated approximate asphalt demand.

For example, suppose that the approximate asphalt demand was calculated to be 6.2 percent. A series of trial mixes might then range from 5.0 to 7.0 percent or from 5.5 to 7.5 percent.

Table A.1 – **Formulas for proportioning materials for recycled hot mixtures**

		For Asphalt Content	
		by wt. of total mix	by wt. of aggregate
% New Asphalt,	P_{nb}	$\dfrac{(100^2 - r\,P_{sb})\,P_b}{100\,(100 - P_{sb})} - \dfrac{(100 - r)\,P_{sb}}{100 - P_{sb}}$	$P_b - \dfrac{(100 - r)\,P_{sb}}{100}$
% RAP,	P_{sm}	$\dfrac{100\,(100 - r)}{100 - P_{sb}} - \dfrac{(100 - r)\,P_b}{100 - P_{sb}}$	$\dfrac{(100 + P_{sb})\,(100 - r)}{100}$
% New Agg. and/or RAM,	P_{ns}	$r - \dfrac{r\,P_b}{100}$	r
Total		100	$100 + P_b$
% New Asphalt to Total Asphalt Content, R		$\dfrac{100\,P_{nb}}{P_b}$	$\dfrac{100\,P_{nb}}{P_b}$

P_{sm} = Percent salvage mix (RAP) in recycled mix
P_b = Asphalt content of recycled mix, %
P_{sb} = Asphalt content of salvaged mix (RAP), %
P_{nb} = Additional asphalt and/or recycling agent in recycled mix, %
P_{ns} = Percent additional aggregate (new or reclaimed aggregate material)
 r   = Percent new and/or reclaimed aggregate material to total aggregate in recycled mix
 R   = Percent new asphalt and/or recycling agent to total asphalt in recycled mix

(3) *Estimated Percent of New Asphalt in Mix*-The quantity of new asphalt to be added to the trial mixes of the recycled mixture, expressed as percent by weight of total mix is calculated by this formula:

$$P_{nb} = \frac{(100^2 - r\,P_{sb})\,P_b}{100\,(100 - P_{sb})} - \frac{(100 - r)\,P_{sb}}{100 - P_{sb}}$$

where: P_{nb}= percent* of new asphalt** in recycled mix
     r = new aggregate (and/or RAM) expressed as a percent* of the total aggregate in the recycled mix
     P_b = percent*, asphalt content of total recycled asphalt mix or asphalt demand, determined by CKE or empirical formula in item (2) above
     P_{sb}= percent*, asphalt content of reclaimed asphalt pavement

* Expressed as a whole number.
**Plus recycling agent, if used.

For example, suppose the asphalt content, P_{sb}, of the RAP is 4.7 percent and r = 75%, then

$$P_{nb} = \frac{(100^2 - 75 \times 4.7)\,P_b}{100\,(100 - 4.7)} - \frac{(100 - 75)\,4.7}{100 - 4.7} = 1.01\,P_b - 1.23$$

The percentages of new asphalt for any asphalt content may now be readily determined.

Note: The formula is for asphalt content expressed as percent by weight of total mix. If asphalt contents are expressed as percent by weight of aggregate the formula for calculating quantity of new asphalt is:

$$P_{nb} = P_b - \frac{(100 - r) P_{sb}}{100}$$

(See Table A.1).

(4) *Select Grade of New Asphalt* - Using Figure A.2, a target viscosity of the asphalt blend is selected. A commonly selected target point is the viscosity at the mid-range of an AC-20 asphalt or 2,000 poises.

The percent of the new asphalt, P_{nb}, to the total asphalt content, P_b, is expressed by this formula:

$$R = \frac{100\, P_{nb}}{P_b}$$

For example, suppose the mix described in Step (3) is to have an estimated asphalt content of 6.2 percent. The amount of new asphalt to be added (from Step 3) is:

$$P_{nb} = 1.01 \times 6.2 - 1.23 = 5.0 \text{ percent}$$

Then:

$$R = \frac{100\,(5.0)}{6.2} = 81 \text{ percent}$$

The grade of new asphalt (and/or recycling agent) is determined using a log-log viscosity versus percent new asphalt blending chart such as Figure A.2. A target viscosity for the blend of recovered asphalt and the new asphalt (and/or recycling agent) is selected. The target viscosity is usually the viscosity of the mid-range of the grade of asphalt normally used depending on type of construction, climatic conditions, amount and nature of traffic.

Plot the viscosity of the aged asphalt in the RAP on the left hand vertical scale, Point A, as illustrated in Figure A.2. Draw a vertical line representing the percentage of new asphalt, R, calculated above and determine its intersection with the horizontal line representing the target viscosity, Point B. Then draw a straight line from Point A, through Point B and extend it to intersect the right hand scale,

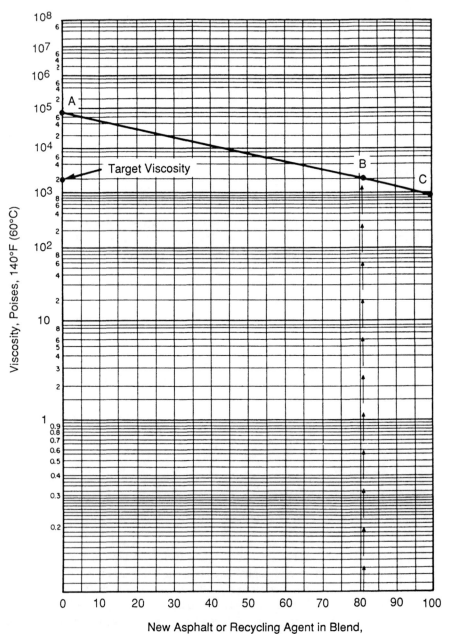

Figure A.2 – **Asphalt viscosity blending chart**

Point C. Point C is the viscosity at 60°C (140°F) of the new asphalt (and/or recycling agent) required to blend with the asphalt in the reclaimed asphalt pavement to obtain the target viscosity in the blend. Select the grade of new asphalt that has a viscosity range that includes or is closet to the viscosity at Point C.

To plot a point using the vertical scale, consider expressing the viscosity using 10 raised to some power. For example, 75,000 poises would be 7.5×10^4. To plot the point on the vertical scale, 7.5 would be interpolated on the scale between 10^4 and 10^5.

Note: It is suggested that when selecting a grade of asphalt cement for recycling that the following guide be used:
    Up to 20% RAP= No change in asphalt grade
    21% RAP or More= Do not change more than one grade
    (i.e. from AC-20 to AC-10)

---

(5) *Trial Mix Design* - Trial mix designs are then made using the Marshall or Hveem apparatus. The formulas shown in Table A.1 are used for proportioning the ingredients: new asphalt, P_{nb}, reclaimed asphalt pavement (RAP), P_{sm}, and new and/or reclaimed aggregate (RAM), P_{ns}.

Keep in mind that if two different aggregate sources are utilized, such as new aggregate and RAM, the percentages of each of these sources must be determined and the total equal P_{ns}. For example the aggregate blend consisted of:

    60% RAM
,   15% New aggregate
    25% RAP aggregate

    then r = 75

If P_{ns} in a trial mix is to be 61.4 percent, then the percent of RAM (in total mix) will be 61.4 x (60/75) = 49.1 percent and the percent new aggregate will be 61.4 x (15/75) = 12.3. The total equals 61.4 percent.

(6) *Select Job-Mix Formula*

Design Example 1

The reclaimed asphalt pavement has an asphalt content of 5.4 percent by weight of total mix. The viscosity of the asphalt recovered from the reclaimed asphalt pavement is 46,000 poises at 60°C (140°F). The grade of asphalt cement normally used is AC-20, and the target viscosity at a temperature of 60°C (140°F) is 2,000 poises. Gradation of RAP, RAM and new aggregate is:

| | Percent Passing | | |
Sieve Size	RAP Agg.	RAM	New Agg.
25.0 mm (1 in.)	100	100	100
19.0 mm (3/4 in.)	98	92	100
9.5 mm (3/8 in.)	85	45	100
4.75 mm (No.4)	65	19	94
2.36 mm (No. 8)	52	5	85
300 μ m (No. 50)	22	1	26
75 μ m (No. 200)	8	0	6

Approximately 30 percent of RAP is to be used in the mix design.

STEP 1 - Combined Aggregates in Recycled Mixture

| | 30% RAP Aggr. | + | 60% RAM Aggr. | + | 10% NEW Aggr. | = | Comb. Aggr. |
Sieve	% Pass.		% Pass.		% Pass.		% Pass.
25.0 mm (1 in.)	[100×0.3=30.0]	+	[100×0.6=60.0]	+	[100×0.1=10.0]	=	100.0
19.0 mm (3/4 in.)	[98×0.3=29.4]	+	[92×0.6=55.2]	+	[100×0.1=10.0]	=	94.6
9.5 mm (3/8 in.)	[85×0.3=25.5]	+	[45×0.6=27.0]	+	[100×0.1=10.0]	=	62.5
4.75 mm (No. 4)	[65×0.3=19.5]	+	[19×0.6=11.4]	+	[94×0.1= 9.4]	=	40.3
2.36 mm (No. 8)	[52×0.3=15.6]	+	[5×0.6= 3.0]	+	[85×0.1= 8.5]	=	27.1
300 μ m (No. 50)	[22×0.3= 6.6]	+	[1×0.6= 0.6]	+	[26×0.1= 2.6]	=	9.8
75 μ m (No. 200)	[8×0.3= 2.4]	+	[0×0.6= 0]	+	[6×0 .1= 0.6]	=	3.0

Then r = 60 +10 = 70

Job Specification:
ASTM D 3515, Table 1
3/4 in. (19mm) Nom.

Sieve Size	Max. Size % Pass.	Combined Aggr. % Pass.
25.0 mm (1 in.)	100	100.0
19.0 mm (3/4 in.)	90-100	94.6
9.5 mm (3/8 in.)	56-80	62.5
4.75 mm (No. 4)	35-65	40.3
2.36 mm (No. 8)	23-49	27.1
300 μ m (No. 50)	5-19	9.8
75 μ m (No. 200)	2-8	3.0

STEP 2 - Approximate Asphalt Demand of Combined Aggregates

$$P = 0.035 + 0.045b + Kc + F$$
$$= 0.035 \times 72.9 + 0.045 \times 24.1 + 0.20 \times 3.0 + 1.0$$
$$= 5.2 \text{ percent}$$

STEP 3 - Estimated Percent of New Asphalt in Mix

$$P_{nb} = \frac{(100^2 - r\,P_{sb})\,P_b}{100\,(100 - P_{sb})} - \frac{(100 - r)\,P_{sb}}{100 - P_{sb}}$$

$$= \frac{(100^2 - 70 \times 5.4)\,P_b}{100\,(100 - 5.4)} - \frac{(100 - 70)\,5.4}{100 - 5.4}$$

$$= 1.02\,P_b - 1.71$$

For an approximate asphalt demand of 5.2 percent:

$$P_{nb} = 1.02\,(5.2) - 1.71 = 3.6 \text{ percent}$$

The percent of new asphalt, P_{nb}, to total asphalt, P_b, will then be:

$$R = \frac{100\,(3.6)}{5.2} = 69 \text{ percent}$$

STEP 4 - Select Grade of New Asphalt

On Figure A.3, Point A is the viscosity of the aged asphalt at 46,000 poises (4.6×10^4). Point B is located from a target viscosity of 2,000 poises (2.0×10^3) and $R = 69$. The projected line from Point A through Point B to Point C indicated that the viscosity of the new asphalt is 7.0×10^2 poises (700).

Since AC-20 is the normal grade of asphalt cement used in the area of construction, climate and traffic, an AC-10 will be chosen for this project. The AC-10 when blended with the aged asphalt in the RAP should result in an AC-20 within acceptable tolerances.

STEP 5 - Trial Mix Design

Using an aggregate blend of 60 percent RAM, 10 percent new aggregate and 30 percent RAP aggregate, trial mixes of different asphalt contents (varying in 0.5 percent increments) are prepared according to standard Marshall or Hveem mix design procedures.

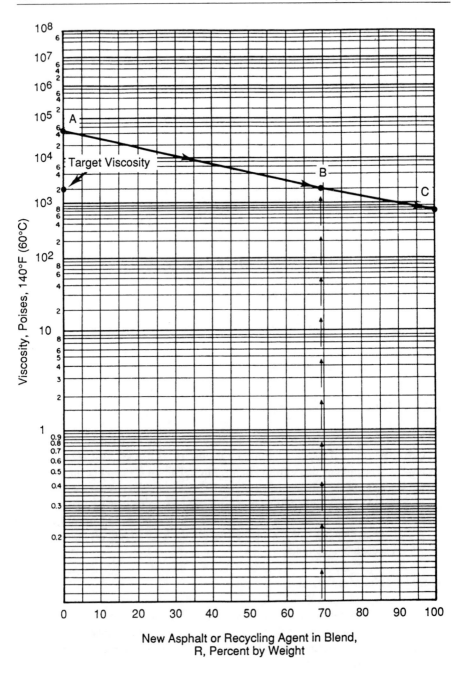

Figure A.3 – **Asphalt viscosity blending chart (Design Example 1)**

The formulas in Table A.1 may be used to determine the percentages of each ingredient in the trial mixes. Since the formula for P_{nb} was calculated in Step 3, the formulas for proportioning the P_{sm} and P_{ns} are:

$$P_{sm} = \frac{100\,(100-r)}{100-P_{sb}} - \frac{(100-r)\,P_b}{100-P_{sb}}$$

$$= \frac{100\,(100-70)}{100-5.4} - \frac{(100-70)\,P_b}{100-5.4}$$

$$= 31.71 - 0.32\,P_b$$

$$P_{ns} = r - \frac{r\,P_b}{100}$$

$$= 70 - \frac{70\,P_b}{100} = 70 - 0.70\,P_b$$

Asphalt Content, P_b	4.5	5.0	5.5	6.0	6.5
$P_{nb} = 1.02\,P_b - 1.71$	2.9	3.4	3.9	4.4	4.9
$P_{sm} = 31.71 - 0.32\,P_b$	30.3	30.1	29.9	29.8	29.6
$P_{ns} = 70 - 0.70\,P_b$	66.8	66.5	66.2	65.8	65.5
Total	100.0	100.0	100.0	100.0	100.0
*% RAM = P_{ns} (60/70)	57.3	57.0	56.7	56.4	56.1
*% New Aggr = P_{ns} (10/70)	9.5	9.5	9.5	9.4	9.4

*The percentages of new aggregate and RAM as a blend were determined as P_{ns}. However, 60 percent RAM and 10 percent new aggregate are to be used in the aggregate blend. The amount of RAM will then be $P_{ns} \times 60/70$ and the new aggregate will be $P_{ns} \times 10/70$.

When preparing trial mixes in the laboratory, it is suggested that the RAP be heated to mixing temperature. The new aggregate and RAM are normally heated to 30°C (50°F) above the mixing temperature. When the aggregate and RAP have been weighed out, dry mixing should begin to thoroughly blend the materials before adding new asphalt. Keeping the RAP at elevated temperatures should be held to a minimum. Otherwise, normal mix design procedures are followed.

STEP 6 - Select Job-Mix Formula

The design, total asphalt content and the mix design are determined according to established standard Marshall or Hveem mix design criteria (as is used for virgin materials).

Design Example 2

Reclaimed asphalt pavement has an asphalt content of 6.0 percent with a viscosity of 100,000 poises. Gradation of RAP, RAM and new aggregate are the same as for Example 1.

STEPS 1 and 2 - Same as Design Example 1.

STEP 3 - Estimated Percent of New Asphalt in Mix

$$P_{nb} = \frac{(100^2 - r\,P_{sb})\,P_b}{100\,(100 - P_{sb})} - \frac{(100 - r)\,P_{sb}}{100 - P_{sb}}$$

$$= \frac{(100^2 - 70 \times 6.0)\,P_b}{100\,(100 - 6.0)} - \frac{(100 - 70)\,6.0}{100 - 6.0}$$

$$= 1.02\,P_b - 1.91$$

For an approximate asphalt demand of 5.2 percent:

$$P_{nb} = 1.02\,(5.2) - 1.91 = 3.4 \text{ percent}$$

STEP 4 - Select Grade of New Asphalt

On Figure A.4, Point A is the viscosity of the aged asphalt at 100,000 poises (1.0×10^5). Point B is located using values of 2,000 poises (2.0×10^3) for target viscosity and $R = 57$, ($100 P_{nb}/P_b = 100 \times 3.4/6.0$) of new asphalt. A line is projected through these two points and intersects the right axis at 1.8×10^2 (180 poises), Point C.

This is a heavily-traveled roadway where the design engineer is concerned with rutting and normally uses an AC-20 in mix design. Figure A.4 can be used to determine how much of a recycling agent to blend with AC-20 to give an apparent viscosity of 180 poises.

Let the AC-20 be the new asphalt and plot 2,000 poises (2.0×10^3) on the left-hand scale, Point D. The viscosity of the recycling agent is 1 poise. Plot this as Point E on the right-hand scale. Connect Points D and E with a straight line. Now determine what percentage, R, of recycling agent will be required to result in a viscosity of 180 poises for the blend. This is plotted as Point F on the line from D to E. The percentage R on the horizontal scale indicates 22 percent. This means that a tank of AC-20 containing 22 percent of the recycling agent should have a viscosity of approximately 180 poises. When this blend is added to the mix for a total asphalt content of about 5.2 percent, the viscosity of the total asphalt in the recycled mix should be 2,000 poises—within acceptable limits.

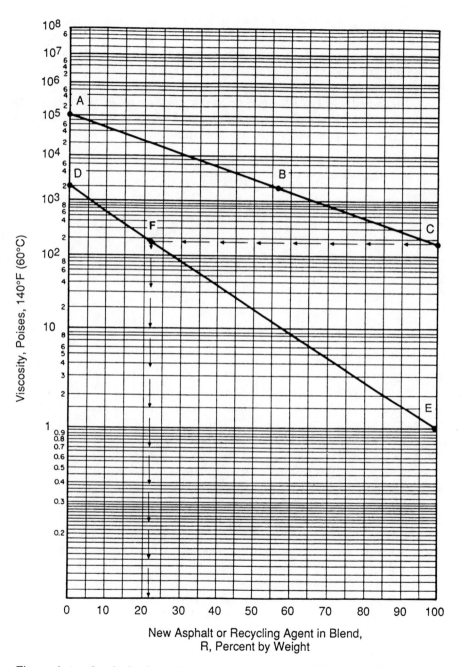

Figure A.4 – **Asphalt viscosity blending chart (Design example 2)**

STEP 5 - Trial Mix Design

Using an aggregate blend of 60 percent RAM, 10 percent new aggregate and 30 percent RAP aggregate, trial mixes of different asphalt contents (varying in 0.5 percent increments on either side of the estimated asphalt demand) are prepared according to standard Marshall or Hveem mix design procedures.

The formulas in Table A.1 may be used to calculate the percentages of each ingredient in the trial mixes. Since the formula for P_{nb} was calculated in Step 3, the formulas for proportioning P_{sm} and P_{ns} are:

$$P_{sm} = \frac{100\,(100 - r)}{100 - P_{sb}} - \frac{(100 - r)\,P_b}{100 - P_{sb}}$$

$$= \frac{100\,(100 - 70)}{100 - 6} - \frac{(100 - 70)\,P_b}{100 - 6}$$

$$= 31.91 - 0.32\,P_b$$

$$P_{ns} = r - \frac{r\,P_b}{100}$$

$$= 70 - \frac{70\,P_b}{100} = 70 - 0.70\,P_b$$

Asphalt Content, P_b	4.0	4.5	5.0	5.5	6.0
$P_{nb} = 1.02\,P_b - 1.91$	2.2	2.7	3.2	3.7	4.2
$P_{sm} = 31.91 - 0.32\,P_b$	30.6	30.5	30.3	30.1	30.0
$P_{ns} = 70 - 0.70\,P_b$	67.2	66.8	66.5	66.2	65.8
Total	100.0	100.0	100.0	100.0	100.0
*% RAM = P_{ns} (60/70)	57.6	57.3	57.0	56.7	56.4
*% New Aggr = P_{ns} (10/70)	9.6	9.5	9.5	9.5	9.4

*The percentages of new aggregate and RAM as a blend were determined as P_{ns}. However, 60 percent RAM and 10 percent new aggregate are to be used in the aggregate blend. The amount of RAM will then be $P_{ns} \times 60/70$ and the new aggregate will be $P_{ns} \times 10/70$.

When preparing trial mixes in the laboratory, it is suggested that the RAP be heated to the mixing temperature. The new aggregate and RAM are normally heated to mixing temperature plus 30°C (50°F). When the aggregate and RAP have been weighed out, dry mixing should begin to thoroughly blend the materials before adding new asphalt. Keeping the RAP at elevated temperatures should be held to a minimum. Otherwise, normal mix design procedures are followed.

STEP 6 - Select Job-Mix Formula

The design, total asphalt content and the mix design are determined according to established standard Marshall or Hveem mix-design criteria (as is used for virgin materials).

Index

M

Marshall, Bruce, 55
Marshall method of mix design, 55-78
  criteria for paving mix, 69
  data preparation, 64
  data, test, trends and relations, 64
  density and voids analysis, 62
  design asphalt content determination, 68
  design criteria, 69
  equipment for, 57, 61
  evaluation of VMA curve, 71
  large aggregate, for, 77
  mix design selection, 69
  mix preparation, 60
  specimen compaction, 61
  stability and flow tests, 62
  stability correlation ratios, 66
  temperatures, 58
  test data interpretation, 68-78
  test property curves, 69
  test report, illustrated, 65
  test specimens preparation, 56
  testing machine, 61, 63
  water bath, 61
Materials control, 17, 120
Maximum specific gravities of mixtures, 50, 114
Mix design, asphalt, 1, 5, 55-107, 123-136
  adjustment in, 12
  durability of pavement, and, 5
  essentiality of, 1
  evaluation of, 5, 12-16, 43-55, 69-77, 100-105
  management of, 16-17
  methods, 1, 55-107
  objectives of, 1, 5
  RAP, using, 123-136
  selection of, 1, 69-77, 100-105
  strength of mix, and, 12
  testing, coordination of, 8-10
  trial mixes, 12-16, 23-24, 125, 131, 136
  verification, 113-121
  warning on, 1, 11, 12

Mix designations, 2, 6-10
Mixer, mechanical, for test mixes, 22
Mixes for tests, preparation of, 20, 60, 89
Mixture, workability of, 5
Mold and hammer in Marshall method, 57
Molds, compaction, in Hveem method, 90

P

Paving mixtures
  analysis of, 43-56
  criteria for, 69, 101
  sample, basic data on, 49
Principles of Construction of Hot Mix Asphalt Pavements (MS-22), 17, 24, 120
Proportioning asphalt and aggregate, 18, 26-42

R

Reclaimed aggregate material (RAM), 123
Random sampling, 17, 117, 120
Reclaimed asphalt pavement (RAP) in mix design, 123-136

S

Sieve analysis, washed, 23, 25
Specific gravity
  aggregate, 37, 48
  definitions, 43
  determination, importance of, 44
  mixture, 47, 50
  VMA determination, and, 45
Specimen testing, 62, 95
Stability
  correlation ratios, 66
  Hveem method, 79, 97-100
  Marshall method, 55, 62, 63
  mix, 5, 11, 13, 16, 35
  voids, and, 13, 16
Stabilometer, Hveem
  calibration of, 105-108
  displacement in, 105
  filling with liquid, 106-107